8. MULTISPECTRAL REMOTE SENSING OVER SEMI-ARID LANDSCAPES FOR RESOURCE MANAGEMENT

Susan L. Ustin and Larry A. Costick

INTRODUCTION

Land Management Problems in Semi-arid Regions

Arid and semi-arid landscapes are fragile ecosystems, sensitive to degradation and desertification. Loss of productivity results from complex interactions between climate impacts and unsustainable land use practices. The United Nations Conference on Environment and Development (UNCED, 1992) estimates 70% of all drylands, totaling about 3.6 B ha or 25% of the Earth's land surface, are affected by desertification. North America and Spain have the largest percentage of deteriorating arid lands, with 90% of arid lands in North America moderately to severely desertified (Dregne, 1986). Dregne (1986) identifies three processes that promote desertification: 1) deterioration of vegetation cover from overgrazing, wood cutting and burning; 2) wind and water erosion from improper land management; and 3) salinization from improper irrigation management. The U.S. National Research Council (1994) identified three criteria for defining rangeland health: soil stability and watershed function, integrity of nutrient cycles and energy flow, and presence of functioning recovery mechanisms. Land managers need new techniques to identify and map locations of environmental problems within their management units to develop effective plans for sustainability. Human activities create disturbances at all spatial scales, making it difficult to develop consistent management decisions, especially over the extended periods necessary to reverse environmental degradation.

The Newton Copper Mine in the California motherlode is typical of abandoned mine operations in the U.S. and globally (Plate 12) and represents a serious restoration and management problem. Pyrite, the primary source of acid drainage is rich in sulfide, releasing heavy metals, *e.g.*, Pb, As, Cd, Ag, and Zn when oxidized. Early identification of erosion and downstream transport would facilitate intervention. Iron and hydroxyls in secondary minerals like jarosite, goethite, and hematite are detectable by visible-infrared sensors, allowing mapping of mine hazards using remote sensing tools. While severe environmental impacts like these are readily identified and observed, less severe but cumulative regional impacts also lead to deterioration of ecosystem functioning. While no habitats are immune to environmental problems, the impact to vegetation and soils is often greater and recovery longer in semi-arid systems.

The Use of Information Technologies in Resource Management

Spatial and information technologies have created a paradigm shift in environmental management by recording location-specific information about habitat conditions that facilitate decision making. Remote sensing technology provides economies of scale for monitoring, analyzing, and in some cases quantifying the parameters needed to characterize natural and human- induced environmental changes. The achievements that have been made in capturing and interpreting remote sensing data in the last three decades have been significant and justify more extensive use of this technology in resource management. Remote sensing technologies can be applied at all scales and resolutions, from hand-held CCD's (charge coupled devices) to sensors mounted on field equipment or aircraft to global mapping satellites. The advent of Geographic Information Systems (GIS), and Global Positioning Satellite (GPS) Systems in the past decade provides a suite of new tools for obtaining spatially explicit environmental information. Use of multi-date imagery makes it practical to quantify environmental conditions and changes at specific sites. Interpretations of change become quantifiable when predictive ecosystem models use spatial and temporal remote sensing data. Methods are evolving as both hardware, software, models and more sophisticated imagery become available.

Remote sensing provides the technology to acquire repeated measurements in a referenced spatial context that produces a synoptic view of the land surface. A wide array of satellites and airborne imaging sensors are available for mapping, monitoring, and extracting information about habitat conditions, and rates and locations of change. Powerful GIS and image processing software programs are now available on PC and MAC platforms in addition to UNIX programs to process this data.

Characteristics of Satellite and Airborne Sensors

Although satellite programs continue to have problems with failed launches, delays, and other engineering failures that have eroded the potential number of land observing satellite systems, still an unprecedented number, approximately 30 non-military sensors, are scheduled for launch between 1999–2002. Stoney (1997a,b) summarized the current and planned land observing systems, which fall into four categories based on their sensor characteristics and resolutions: multispectral systems, hyperspectral systems, hyperspatial systems, and radar/microwave systems. Hyperlinked lists of earth observing sensors and their characteristics are available at many web sites *e.g.*, http://http2.brunel.ac.uk:8080/depts/geo/satmarks/Sensor.html. With such a large number of sensors, and the continually changing launch dates, priorities, and economics, it is difficult to keep track of specific systems and their properties. In addition to satellites, the growing number commercial aircraft sensors, make it possible to obtain high spatial resolution data for virtually any type of spectral resolution sensor. Therefore, in contrast to recent conditions when managers had few choices in use of remote sensing data, users must now identify the sensor best suited for their specific environmental problems.

Multispectral remote sensing systems

The Landsat Multispectral Scanner (MSS) and Thematic Mapper (TM), the Système Pour l'Observation de la Terre (SPOT) with High Resolution Visible (HRV) and panchromatic sensors, and the Indian Remote Sensing (IRS) sensors, have four to eight spectral bands and moderate spatial resolutions (10–30 m), yielding broad area coverage (Table 1). Current plans indicate there will be 13 different satellite sensors having Landsat-like multispectral capabilities by 2002. At least one commercial company, Resource 21, has

Table 1. Summary of remote sensing satellites for earth observation that are scheduled for launch in the 1999–2002 time frame.

Country public/commercial	Satellite	Sensor Name	Resolution in meters — Thematic Mapper Bands							
			PAN	VNIR 1	2	3	4	SWIR 5	7	TIR 6
Frequent Global Coverage, LANDSAT like classification ability										
India Gov	IRS-1 C,D	LISS-3, PAN, (WIFS)	6		23	23	23	70		
Japan Gov	ADEOS	AVNIR	8	16	16	16	16			
China-Brazil Gov	CBERS	CCD, IRMSS	20,80	20	20	20	20	80	80	160
France Gov	Spot 4	HR VIR, (vegetation)			20	20	20	20		
India Gov	IRS-P5	LISS 4, LISS-3			<10	<10	<10	70		
US Gov	Landsat 7	ETM+	15	30	30	30	30	30	30	60
US Com	Resource 21	R21,A-D		10	10	10	10	20		
India Gov	IRS-2A	LISS 2, 3, 4'			5	5	5	70		
High resolution, small area coverage (PAN & VNIR only)										
US Com	EarthWatch	EarlyBird (failed)	3	15	15	15	15			
US Com	SpaceImaging	IKONOS	1	4	4	4	4			
US Com	EarthWatch	QuickBird	1	4	4	4	4			
US Com	Orbimage	OrbView	1,2	8	8	8	8			
US Com	GDF	XXX	2.5							
India Gov	IRS-P6	PAN	1							
Hyperspectral, pixel area coverage like Landsat										
US/Japan Gov	Terra	Aster			15	15	15	6 bands @30		5@90
US Gov	EO-1	Hyperion		128 bands @30				256 bands @30		
US Com	Orbimage	Orbview-4	1	Minimum of 200 bands from 0.4 to 5 um@<10						
US Com	STDC	NEMO	1	200 bands from 0.4 to 2.5um						
Australia/Com		AIRES-1	10	32 bands 400–1100nm and 32 bands 2000–2500nm @ 30m						

IR	Infrared spectral region	All satellites in polar sun synchonous orbits except
VNIR	Visible and near IR	Quickbird.
SWIR	Short wave IR	Stereo satellites have 2–3 day site revisit potential.
TIR	Thermal IR	

announced it will launch four copies of a TM-like satellite in order to achieve worldwide weekly or better repeat coverage. These satellites all have three visible bands (blue, green, red) and one near-infrared band. In addition, all but the Japanese AVNIR sensor have at least one shortwave-infrared (SWIR) band (in the 1.5–2.5 μm range), and two have a second SWIR band and a thermal infrared (TIR) band. Several of these satellites include a high spatial resolution panchromatic band having 5–15 m resolution, while multispectral bands have spatial resolutions between 15–100 m. These sensors are optimal for spatially explicit mapping and analysis of watersheds and regional processes, observation of land use and land cover classes at the landscape scale, mapping of roads and infrastructure, estimates of vegetative cover, measures of fragmentation, and spatial patterns related to natural disturbances, *e.g.*, severe storms, wildfires, floods, erosion, and droughts. Vegetation patterns related to clear-cut logging, forest stands of different structure and species mixture classes, agricultural patterns, and mining activity are generally at a scale suitable for mapping with these sensors.

Hyperspectral imaging systems

This technology is relatively new and represents the most advanced imaging systems. These sensors have large numbers (50–1000's) of adjacent narrow wavelength spectral bands so that the band sequence creates a high resolution spectrum for each pixel. These sensors are also called Imaging Spectrometers, conveying the concept that the two-dimensional spatial array of pixels for each spectral band forms an image. Because of the large number of bands, a dataset is often described as an "image cube" having both spatial (w,y) and spectral (z) dimensions for all pixels. NASA plans to launch the first hyperspectral satellite, Hyperion on the EO-1 platform in summer 2000. Three commercial hyperspectral sensors are expected to be launched into orbit in the 2000–2001 time frame. These are: OrbView-4 (Orbital Sciences Corporation), Navy Earth Mapping Observer (NEMO), operated commercially by Applied Technology Development Corporation (ATDC), and Australian Resource Information and Environment Satellite (AIRES), a joint Australian commercial-government venture. Each of these sensors has spectral coverage in about 200 channels between 0.40–2.50 mm and pixel resolutions of 30 m or better. More than 30 airborne hyperspectral sensors have been operated in various countries (for a list, see http://rst.gsfc.nasa.gov/sect13/is–list.html) with increased availability expected in the future.

At the expense of adding significantly more data to analyze, these sensors are capable of identifying many specific biogeochemical conditions in plants, soils, and geologic materials based on their spectral signatures. Among the properties that can be identified are many specific minerals, clay minerals, carbonates, organic matter, and iron in soil (Baumgardner *et al.*, 1985; Clark, 1999; Ben-dor *et al.*, 1999) and plant components, *e.g.*, pigments, water, and structural carbon compounds like cellulose (Ustin *et al.*, 1999). With this spectral information it is possible to make inferences about the state of health of the ecosystem and possibly, the presence of stress agents, like contaminants.

Hyperspatial imaging sensors

High resolution data is useful to identify specific impacts affecting only localized conditions. Twelve satellites capable of measuring very high spatial resolution imagery from space are planned for launch in the 2000 time frame (Stoney, 1997a,b). These commercial sensors will provide images at spatial resolutions similar to that currently provided by the airborne photogrammetry industry. The QuickBird (EarthWatch Spacecraft), OrbView-3 (Orbital Sciences Coroporation), and Ikonos (Space Imaging Corporation — launched 24 September 1999) satellites, will have a 1 m resolution panchromatic band sensor and 4m resolution, four-band (blue, green, red, and NIR) multispectral sensor on board. Because of the narrow ground swaths (strips only 4–36 km in width), global repeat coverage will require periods from 4 months to 2 years (1997a). Off nadir side-to-side and fore-to-aft pointing capability will increase the frequency of repeat site measurements. Pointing capabilities also allow acquisition of stereo images, making it possible to obtain detailed high-resolution digital elevation maps and estimates of canopy and surface heights. While scheduling flexibility is lost compared to digital airborne photography, satellite imagery provides advantages by reducing processing time to co-register and mosaic multiple air photo images. The stability of the space platform reduces spatial distortions characteristically present in airborne data due to aircraft motion and the digital spectral data adds information. Sun-synchronous orbits reduce the variability introduced by changing sun elevation angles with time-of-day. Therefore considerable savings in analyst time is likely. Virtually any application where aerial photography is currently used is a candidate for using this type of sensor data. An extensive literature is available for resource management applications, therefore the following sections are intended to illustrate typical uses of these different types of sensors and are not comprehensive.

APPLICATIONS OF MULTISPECTRAL, HYPERSPECTRAL AND HYPERSPATIAL SENSOR DATA FOR RESOURCE MANAGEMENT

Multispectral Mapping of Mine Locations and Natural Hazards

Regional synoptic views aid assessment of the relative hazards of mines, their proximity to habitation, and the extent and location relative to the location on the original mining claim. California and U.S. agencies have mine location databases but these are not always accurate. The use of multispectral satellite data, like the Landsat TM can improve the correct identification of mine locations and with GIS database information, map the aerial extent of mining activities. Because mine activity typically involves removal and loss of vegetation cover, the first stage of image analysis identifies sites where plant cover is anomalously low. Plate 12 (bottom) shows a false color image using Landsat TM bands 4, 3, and 2 (near-infrared (NIR), red and green spectral bands) displayed as red, green, blue, respectively, of Central California's Sierra Nevada foothills and motherlode region. Montane forests and riparian vegetation are seen in red tones, grasslands which are dry in summer, are seen in shades of blue and green, while bare soils are whiter, because of the higher NIR reflectance relative to plant litter. The spatial resolution can not locate individual shafts but is adequate to map mine spoils, mill waste, open pits, and hydraulic cuts when images are displayed at full resolution.

Many studies have investigated hydrological indicators of desertification, with most focusing on using spatial patterns in vegetation density and abundance. Smith *et al.* (1990b) used a summer TM image to estimate vegetation cover using spectral unmixing of all six bands, TM thermal temperatures, and a DEM to examine regional water use patterns among semiarid vegetation communities in Owens Valley, California, in the rainshadow of the Sierra Nevada Range. At a much larger scale, Tucker *et al.* (1991) used vegetation abundance, measured by the Advanced Very High Resolution Radiometer (AVHRR) meteorological satellite, to map interannual variation in the extent of the Sahara Desert from 1980 to 1990. Their analysis used a bimonthly time series of the two-band green vegetation index to follow the response of desert vegetation to precipitation and drought and demonstrated interannual changes in extent of the arid desert of more than 1.3 M km^2.

Hyperspectral Biogeochemical Detection and Land Cover Mapping

It is generally difficult to determine the "state of health" of vegetation using multispectral remote sensing but some new hyperspectral sensors and techniques show considerable promise. Many plant compounds have been identified using spectroscopic assays (*e.g.*, Curran and Kupiec, 1995; Dawson *et al.*, 1997; Jacquemoud *et al.*, 1996; McLellan *et al.*, 1991), suggesting that spectral characteristics can be used to measure biogeochemical conditions and ecosystem "health." Because matter absorbs and emits energy at specific wavelengths that is determined by the chemical composition and structure of the compounds, it is possible to link physical conditions to image properties. In some cases, determination of general classes, *e.g.*, green foliage, leaf litter and standing litter are sufficient to determine canopy condition.

Spectral Mixture Analysis (SMA, Adams *et al.*, 1986; Smith *et al.*, 1990a,b) is a flexible analysis for a variety of applications, can be used with multispectral and hyperspectral data, and has been shown to provide a constant frame of reference from which to make quantitative interpretations of biophysical changes over space and time (Adams *et al.*, 1995). SMA has been widely adopted as the method of choice for hyperspectral analysis, however as analyses have attempted more quantitative analysis of minor scene constituents, non-linear methods are needed (Mustard and Sunshine, 1999; Pinzon *et al.*, 1998; Ray and Murray, 1996). For some minerals and other materials, identification can be based on spectral feature fitting (Zhang *et al.*, 1996; Sanderson *et al.*, 1998; Clark, 1999). Multiple endmember analysis (Roberts *et al.*, 1998) attempts to map individual plant species or community types based on spectral matching. Hierarchical Foreground/Background Analysis (Pinzon *et al.*, 1998; Ustin *et al.*, 1998; Palacios-Orueta *et al.*, 1999) uses a non-linear image-processing approach that extracts minor spectral contributions from overall scene variance, *e.g.*, variation in biochemical composition, using hyperspectral sensor data (Mustard and Sunshine, 1999).

Mapping California's oak woodland savannas

Semi-arid grasslands and scrublands are dormant in the dry season and detection must be keyed to plant litter instead of green foliage (Roberts *et al.*, 1993; Ustin 1992). Ustin *et al.* (1996) attempted to quantify dry grass residues in semi-arid grasslands of the Central Valley of California. The residual litter at the end of the summer is an indicator of rangeland degradation and excessive grazing, accumulated litter predicts wildfire risk, while loss of cover predicts soil erosion hazard. Figure 1 shows the relative concentrations (gray scale from lowest to highest) of green foliage, litter, and soil from a region of the inner Coast Range of California (see Ustin *et al.*, 1996). These results are derived from a SMA using 1992 NASA 224 band Advanced Visible Infrared Imaging Spectrometer (AVIRIS) data. Areas that have high concentrations of plant litter are primarily located on the south and west facing slopes. The relative proportions of the endmembers provides a good classifier to map savanna communities in this region (Ustin *et al.*, 1996). The reference spectra used as endmembers for the SMA analysis are shown in Figure 1. The spectral differences between plant litter and soils permit separation, although misclassification of the quantities remained significant. Errors are attributable to a combination of inappropriate scale of field sampling, errors in calibrating the data to surface reflectance and low sensor signal-to-noise ratio (SNR). As sensor technologies have become more stable and radiometric

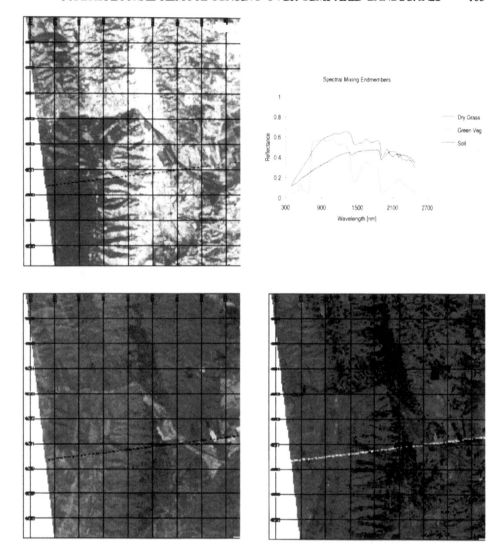

Figure 1. Gray scale endmember fraction images for dry grass, green vegetation and soil. The AVIRIS data was analyzed using linear SMA over an area of the central Coast Range in California. The grid shows UTM coordinates for the site at 1 km intervals. The gray scale goes from black (0%) to white (100%) pixel reflectance as modelled by the endmember. Dry grass residues are very high along the inner slopes of the coast range and on harvested fields. Some cropland and riparian zones have high green fractions and few pixels have high exposures of bare soil.

resolution has increased, the inadequacy of field data to validate analyses has become a more critical problem. Recent estimates suggest a SNR of 200:1 (for a 30% reflectance target, at 30° zenith angle) is necessary to accurately identify and quantify absorption features in the SWIR region. The biochemical concentrations of nitrogen and lignin are of ecological interest and efforts have been made to measure them using hyperspectral sensors. Kokaly and Clark (1998), estimate an SNR of 700 is required in the SWIR region for a precision of 0.5% nitrogen (by dry weight). Because foliar nitrogen is generally less then 3% dry weight, this requirement would require an unrealizable SNR of >5000. Canopy lignin estimates will require a SNR of 500 at 50% reflectance to achieve 5% precision. The first generation of hyperspectral satellites will not have SNR's of this resolution.

Locating hazardous mine wastes

Techniques to extract mineral information from hyperspectral sensors have ranged from simple false color composites to more sophisticated mathematical transformations like Principal Components Analysis and spectral matching. Generally, the approaches used for hyperspectral analysis are more deterministic than the statistical methods employed for multispectral analysis. Monitoring and evaluation of geological conditions and hazardous wastes almost inevitably requires some assessment of vegetation condition or evidence of land cover disturbance, typically either observing vegetative dieback, poor vegetative growth, or bare zones (Ustin *et al.*, 1999). Thus, knowledge of the ecological and physiological constraints is essential for all but the most arid applications. Many mines, mineral outcrops or contaminant sources are small point-sources located within a larger geologic and ecological context.

Because many minerals have unique spectra, it is possible to identify the types of contaminants in bare soils. An extensive reference library of mineral spectra is available from the USGS Digital Spectral Library (Clark *et al.*, 1993; Clark, 1999) at http://speclab.cr.usgs.gov and a library of more than 2000 spectra is available from the ASTER web site http://speclib.jpl.nasa.gov maintained by Simon Hook of the Jet Propulsion Laboratory. This library includes the spectra compiled by Salisbury *et al.* (1991). The most extensive spectral library of soils from many soil series across the world are available from E. Stoner at Purdue University.

Examples of reference spectra used to identify minerals in hyperspectral image data are shown in Figure 2. Narrow band spectral features significantly contribute to identification of minerals based on wavelength position and band depth, and overall albedo. Reference spectra like these are used in spectral matching algorithms and other feature detection algorithms (Mustard and Sunshine, 1999). Swayze *et al.* (1996) used AVIRIS to map heavy metals and produced an iron-bearing mineral map for the Leadville mine site in Colorado (130 km SW Denver). This mine site is one of the most seriously contaminated in the United States and analysis shows that significant off-site transport has occurred, where contamination is located, and what geologic minerals are present.

Hyperspatial Mapping of Landscape Structure

Soil salinity and soil quality

When spatial structure is complex and variable at a fine grain size, only high resolution images can capture the patterns. One example of this management problem is the saline and carbonate soils on the western side of the San Joaquin Valley of California, which are derived from Cretaceous marine sediments. While growers have improved management, adopted salt tolerant crops, and improved reclamation practices, the productivity in the Valley dropped 10% ($32.3 million per year) since 1970 because of high salinity.

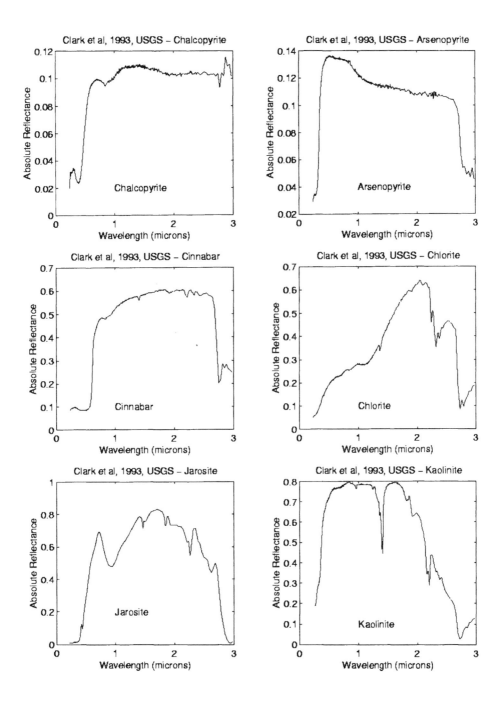

Figure 2. Six representative geologic spectra of minerals associated with mine spoils. The unique spectra of minerals often enables their identification. Data is from the USGS Speclab database.

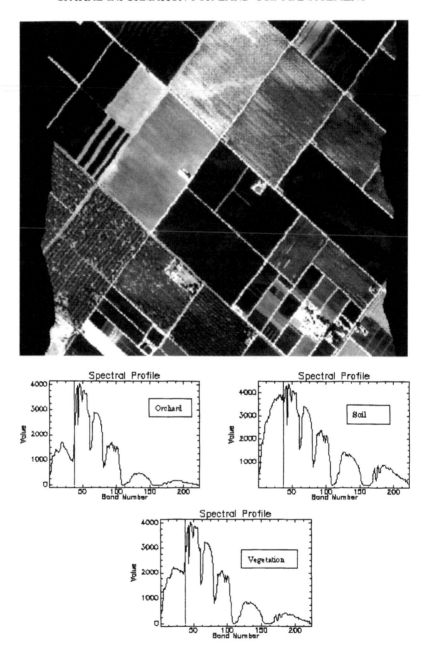

Figure 3. High spatial resolution AVIRIS image acquired in October 1998 of agricultural fields in the San Joaquin Valley, California that are affected by soil salinity problems. Data has spatial resolution of about 4m and the image is 612 pixels wide and 518 pixels long. The image is a single NIR band in a gray scale display. The spectra above show three pixel spectra obtained from the orchard, a bare soil plot and a row crop vegetation plot. Data have been corrected for aircraft motion and atmospherically corrected (using a modified MODTRAN III radiative transfer model) to surface reflectance. Agricultural pixels are partially senescent accounting for the high visible reflectance.

Figure 3 shows a single NIR band high spatial (ca. 4 m pixel) resolution AVIRIS image from October, 1998. The figure clearly shows texture related soil patterns and poor crop development from salinity. The full spectrum of each pixel allows mapping of soil salinity at a scale suitable for irrigation management.

Vegetation management and encroachment on economic resources

Preventing vegetation encroachment is an important part of disaster preparedness and hazard reduction for all power, telecommunications, railroads, and pipeline utilities. The cost of remediation is a significant operating cost and efficiency is essential. Remote sensing images can be used to reduce risk and costs for vegetation treatments, when high spatial resolution data are used to monitor the typically narrow management corridors. Typically, monitoring of wildfire risk, windthrows, and other hazards has been done by sight (*e.g.*, driving along corridors) or by analysis of aerial photos. The linear extent of utility corridors and the need for repeated observations makes use of digital imagery attractive, since digital aerial cameras and commercial satellite sensors can obtain 1 m resolution or better image data. High spatial resolution images impose the same data volume constraints as hyperspectral data. However, for applications only a narrow swath is needed, thus reducing data volume. If a larger regional context is needed, hyperspatial images along the corridor can be fused with lower resolution image data.

Figure 4 (top) shows a 1 m resolution, 4–band airborne ADAR-5500 (Positive Systems, Inc., Whitefish, MT) image, of the low elevation Lake Tabeaud in the Sierra Nevada foothills (Figure 1). In this region, the landscape is composed of discontinuous mixed hardwood and conifer forests. In the enlarged image (NIR band), the 2 cm diameter powerlines (160 kb pole line) are observed crossing the right arm of the lake (Figure 4 middle). This figure illustrates the detection of subpixel features when there is sufficient spectral contrast with the surroundings. In addition, it demonstrates that hyperspatial data often allows visual observation to characterize the images. The photo (Figure 4 bottom) looks towards the northern bank and provides perspective on the scale of the powerlines and the clearings. Vegetation has been cleared from the powerline corridors on both banks and boundaries are easily visible; the break in slope is a water canal which is seen on the image as dark linear feature.

The width of the corridor can be estimated over any length segment and identify sites where vegetation mitigation is needed. Data acquired after mitigation can be used for quality control and compliance. Vegetation and land cover type maps can be classified using the multispectral data. Many land uses that require management intervention can be observed in the image, *e.g.*, soil deposition in the lake, logging and other land uses, condition of roads, and locations of localized erosion. Photogrammetric quality stereo digital imagery can be used to estimate canopy or tower heights although today's commercial software is inadequate to automatically pair stereo data; manual measurements can be done using photographic procedures.

Predicting Future Landscape Conditions Using Ecosystem Models

Accurate indexing methodologies are needed to identify cumulative impacts that result from repeated logging and grazing activities. Cumulative watershed effects (CWE) are the

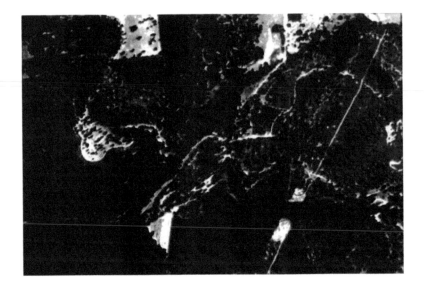

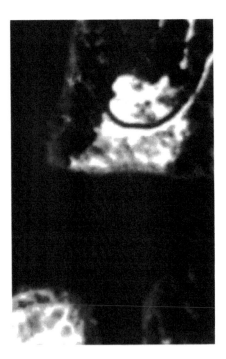

Figure 4. High spatial resolution (1 m) ADAR 5500 data of the Lake Tabeaud region in the foothills of the central Sierra Nevada Range, California. The upper image shows a composite image of the lake and the surrounding region. The lower left image is an enlargement of the powerline segment crossing the lake and showing the subpixel detection of the power lines (dark lines are shadows, light are the powerlines). The photo provides perspective on the south facing slope, on the north shore of the lake.

synergistic impacts caused by the sum of all natural and anthropogenic actions (Berg *et al.*, 1996; Costick 1996). The CWE approach helps managers and regulators objectively assess impacts. Remote sensing driven indexing models with spatially explicit accounting systems identify sites at greater risk of producing accelerated erosion or sedimentation. The CWE index used a DEM to estimate slope steepness, a soil database to identify easily detachable soil types (*i.e.*, have a high *k* factor), and Landsat TM data to identify locations of bare (non-vegetation covered) soils. The normalized ratio of the red and NIR bands (band 4 – band 3 / band 4 + band 3) was used to identify sites having low vegetation cover (Plate 13 top). Plate 13 (bottom) shows bare soils as blue-to white areas, conifer forests as dark red, meadows and broadleaf riparian vegetation as light red. The oval area of lighter-red covering the right half of the scene, north and south of the road is a wildfire scar. The erosion potential of a site is assumed to increase if 2 or 3 CWE risk factors are present. Plate 13 (bottom) shows that the CWE risk of mass wasting is critical in only a small part of the region. This information allows managers to effectively schedule maintenance and to objectively evaluate different land use requests. The South Fork of the American River in the Eldorado National Forest (Plate 13 bottom) is one of the few areas in the Sierra Nevada Range to exceed all three risk parameter thresholds. Plate 13 (bottom) shows where three major landslides occurred in 1983 and 1999 that closed a major transcontinental highway for many months. Several other potential landslide sites are being actively monitored and considerable mitigation has been done along the highway corridor to prevent further problems.

CONCLUSIONS

The expanded use of remote sensing for resource monitoring is promising. The dramatic increase in availability of digital remotely sensed data on airborne and spaceborne platforms provides an opportunity for resource managers to enhance spatially explicit management by incorporating specific information about site conditions. Virtually all parts of the electromagnetic spectrum that are transmitted through the atmosphere are now or soon will be measured by a public domain or commercial sensor. Hand-carried or truck mounted sensors, towers and other platforms make it possible to acquire data at all spatial resolutions and facilitate the continued development of interpretations based on physical absorption and scattering processes. The parallel development of related technologies like GPS and GIS and computer hardware, analysis, and display capability enhance the usefulness of remote sensing data and lead toward developing modeling and prediction methodologies that are tailored to specific sites.

REFERENCES

1. J.B. Adams, M.O. Smith and P.E. Johnson (1986) *J. Geophys. Res.*, **91**, 8098–8112.
2. J.B. Adams, D.E. Sabol, V. Kapos, R. Almeida Filho, D.A. Roberts, M.O. Smith and A.R. Gillespie (1995) *Remote Sens. Environ.*, **52**, 137–154.
3. M.F. Baumgardner, L.F. Silva, L.L. Biehl and E.R. Stoner (1985) *Adv. Agron.*, **38**, 1–44.

4.   E. Ben-dor, J.R. Irons and G. Epema (1999) Soil reflectance, in *Manual of Remote Sensing: for the Earth Sciences*, edited by A. Rencz, Vol. 3, pp. 111–188. (Bethesda, Maryland: American Society of Photogrammetry and Remote Sensing).

5.   N.H. Berg, K.B. Roby and B.J. McGurk (1996) Cumulative watershed effects: Applicability of available methodologies to the Sierra Nevada, in Status of the Sierra Nevada, Assessments, Commissioned Reports and Background Information, Sierra Nevada Ecosystem Project, Final Report to Congress, *Wildland Resource Center Report No. 38*, University of California Davis, **3**, 39–78.

6.   R.N. Clark, G.A. Swayze, A. Gallagher, T.V.V. King and W.M. Calvin (1993) *The U.S. Geological Survey, Digital Spectral Library*, Ver. 1: 0.2–3.0 mm, USGS Open File Rep. 93–592, U.S. Geological Survey, Washington, D.C., 1326 pp.

7.   R.N. Clark (1999) Spectroscopy of rocks and minerals and principles of spectroscopy, in *Manual of Remote Sensing: for the Earth Sciences*, edited by A. Rencz, Vol. 3, pp. 3–38. (Bethesda, Maryland: American Society of Photogrammetry and Remote Sensing).

8.   L.A. Costick (1996) Indexing current watershed conditions using remote sensing and GIS, in Status of the Sierra Nevada, Assessments, Commissioned Reports and Background Information, Sierra Nevada Ecosystem Project, Final Report to Congress, *Wildland Resource Center Report No. 38*, University of California Davis, **3**, 79–152.

9.   P.J. Curran and J.A. Kupiec (1995) Imaging spectrometry: A new tool for ecology, in *Advances in Environmental Remote Sensing*, edited by F.M. Danson and E.M. Plummer, pp. 71–88. (Chichester: Wiley).

10.   H.W. Cutforth and B.G. McConkey (1997) *Can. J. Plant Sci.*, **77**, 359–366.

11.   T.P. Dawson, P.J. Curran and S.E. Plummer (1997) The potential for understanding the biochemical signal in the spectra of forest canopies using a coupled leaf and canopy model, in *Physical Measurements and Signatures in Remote Sensing*, edited by A. Guyot and T. Phulpin, pp. 463–470. (Rotterdam: Balkema).

12.   H.E. Dregne (1986) Desertification of arid lands, in *Physics of desertification*, edited by F. El-Baz and M.H.A. Hassan. (Dordrecht, The Netherlands: Martinus, Nijhoff).

13.   S. Jacquemoud, S.L. Ustin, J. Verdebout, G. Schmuck, G. Andreoli and B. Hosgood (1996) *Remote Sens. Environ.*, **56**, 194–202.

14.   R.F. Kokaly and R.N. Clark (1998) Spectroscopic determination of leaf biochemisty: use of normalized absorption band depths and laboratory measurements and possible extension to remote sensing, Summ. 7th Ann. Jet Propulsion Laboratory Airborne Science Workshop, January 12–16, JPL Publ. 97–21, **1**, 235–244.

15.   T.M. McLellan, J.D. Aber, M.E. Martin, J.M. Melillo, K.J. Nadelhoffer (1991) *Can. J. Forest Res.*, **21**, 1684–1688.

16.   J.F. Mustard and J.M. Sunshine (1999) Spectral analysis for earth science investigation, *in Manual of Remote Sensing: for the Earth Sciences*, edited by A. Rencz, Vol. 3 pp. 249–250. (Bethesda, Maryland: American Society of Photogrammetry and Remote Sensing).

17.   Palacios-Orueta, J.E. Pinzon and S.L. Ustin (1999) *Remote Sens. Environ.*, **68**, 138–151.

18.   J.E. Pinzon, S.L. Ustin, C.M. Castaneda and M.O. Smith (1998) *IEEE Trans. Geosci. Remote Sens.*, **36**, 1–15.

19.   T.W. Ray and B.C. Murray (1996) *Remote Sens. Environ.*, **55**, 59–64.

20.   D.A. Roberts, M.O. Smith and J.B. Adams (1993) Green vegetation, nonphotosynthetic vegetation and soils in AVIRIS data. *Remote Sens. Environ.*, **44**, 255–269.

21.   D.A. Roberts, M. Gardner, R. Church, S.L. Ustin, G. Scheer and R.O. Green (1998) *Remote Sens. Environ.*, **65**, 267–279.

22.   E.W. Sanderson, M. Zhang, S.L. Ustin and E. Rejmankova (1998) *Landscape Ecol.*, **13**, 79–92.

23. J.W. Salisbury, L.S. Walter, N. Vergo and D.M. D'Aria (1991) *Infrared (2.1–2.5 mm) Spectra of Minerals.* (Baltimore: Johns Hopkins University Press), 267 pp.
24. M.O. Smith, S.L. Ustin, J.B. Adams and A.R. Gillespie (1990a) *Remote Sens. Environ.*, **29**, 1–26.
25. M.O. Smith, S.L. Ustin, J.B. Adams and A.R. Gillespie (1990b) *Remote Sens. Environ.*, **29**, 27–52.
26. W.E. Stoney (1997a) Land sensing satellites in the year 2000, in *IGARSS'97, Proceedings of the International Geoscience and Remote Sensing Symposium, Singapore*, edited by T. L. Stein (IEEE Geoscience and Remote Sensing Society, Texas, 1997a) (http://geo.arc.nasa.gov/sge/landsat/pecora.html).
27. W.E. Stoney (1997b) *Geotimes*, **42**, 18.
28. G.A. Swayze, R.N. Clark, R.M. Pearson and K.E. Livo (1996) Mapping acid-generating minerals at the California gulch superfund site in Leadville, Colorado using imaging spectrometry, Summ. 6th Ann. Jet Propulsion Laboratory Airborne Science Workshop, March 4–8, JPL Publ. 96–4, **1**, 231–234.
29. J. Tucker, H.E. Dregne and W.W. Newcomb (1991) *Science*, **253**, 299–301.
30. UNCED (1992) Chapter 12 of Agenda 21, Report of the United Nations Conference on Environment and Development, Rio de Janeiro, 3–13 June.
31. U.S. National Research Council (1994) *Rangeland Health: New Methods to Classify, Inventory and Monitor Rangelands.* (National Academy Press, Washington DC, 1994), 182 p.
32. S.L. Ustin (1992) Thematic Mapper vegetation cover model, in *Evapotranspiration measurements of native vegetation, Owens Valley, California, June 1986*, edited by D.H. Wilson, R.J. Reginato and K.J. Hollett, U.S.G.S. Water-Resources Investigations Report 91–4159, pp. 61–70.
33. S.L. Ustin, M.O. Smith and J.B. Adams (1993) Remote Sensing of Ecological Processes: A strategy for Developing Ecological Models Using Spectral Mixture Analysis, in *Scaling Physiological Processes: Leaf to Globe*, edited by J. Ehlringer and C. Field, pp. 339–357. (New York: Academic Press).
34. S.L. Ustin, Q.J. Hart, L. Duan and G. Scheer (1996) *Int. J. Remote Sens.*, **17**, 3015–3036.
35. S.L. Ustin, D.A. Roberts and Q.J. Hart, Seasonal Vegetation Patterns in a California Coastal Savanna Derived from Advanced Visible/Infrared Imaging Spectrometer (AVIRIS) Data, in *Remote Sensing Change Detection: Environmental Monitoring Applications and Methods*, edited by C.D. Elvidge and R. Lunetta, pp. 163–180. (MI: Ann Arbor Press).
36. S.L. Ustin, D.A. Roberts, S. Jacquemoud, J. Pinzon, M. Gardner, G. Scheer, C.M. Castaneda and A. Palacios (1998) *Remote Sens. Environ.*, **65**, 280–291.
37. S.L. Ustin, M.O. Smith, S. Jacquemoud, M.M. Verstraete and Y. Govaerts (1999) Geobotany: Vegetation Mapping in Earth Sciences, in *Manual of Remote Sensing: for the Earth Sciences*, edited by A. Rencz, Vol. 3, 189–248. (Bethesda, Maryland: American Society of Photogrammetry and Remote Sensing).
38. M. Zhang, S.L. Ustin, E. Rejmankova and E.W. Sanderson (1996) *Ecol. Appl.* **7**, 1039–1053.

9. APPLICATIONS FOR SPATIAL DATA IN GRASSLAND MONITORING AND MANAGEMENT

Michael J. Hill

INTRODUCTION

This chapter deals with the development of layers of spatial data which provide quantitative or qualitative measures of grassland properties and their integration into a decision-making framework. Grassland provides special difficulties in the application of spatial data. In extensive systems, variation in biophysical characteristics of grassland vegetation is continuous and heterogeneous. Management and grazing modify the growth of grassland in an irregular time series of events which are spatially asynchronous. Nevertheless, spatial data may be applied to management issues for grassland at scales from continental to regional to property, paddocks and finally to management units within paddocks. The concept of applying spatial data to management of grazing lands was first clearly elucidated in Australia by Graetz *et al.* (1984) when describing the use of Landsat satellite data to classify rangeland types, derive vegetation indices and detect, quantify and interpret change. In this chapter, I will use the work carried out by myself and my colleagues in Australia to illustrate some of the developments in spatial data as an information source for grassland management. This work was carried out in the context of the grassland and grazing systems of the southern temperate and Mediterranean parts of Australia.

Pastures and Grassland in Southern Australia

The grassland environment of Australia provides a unique set of characteristics which influence the development and application of spatial data for management decisions. Native perennial grasslands growing on impoverished soils in relatively arid conditions have been heavily modified or replaced by introduction of exotic European *Trifolium*, *Medicago* and other leguminous species through widespread application of fertilizer containing phosphorus and sulfur. With nitrogen fixation by clover raising fertility levels, nitrogen-loving exotic grasses were also introduced. Native species are adapted to a lower fertility domain and a less intensive grazing regime than the introduced exotic species. Extensive sheep and cattle systems resulted in highly selective grazing of heterogeneous mixtures of cool season and warm season native perennial grasses leading to dominance of these grasslands by less palatable and lower quality species. The most successful introduced species is a cool season annual *Trifolium subterraneum*. The widespread naturalization of these annuals results in a change in the growth pattern of many grassland areas from opportunistic or aseasonal growth to highly seasonal growth in winter and spring.

Thus development of grassland in Australia may be represented by a fertility transition from low levels to high levels of sulfur, phosphorus and nitrogen paralleled by a change

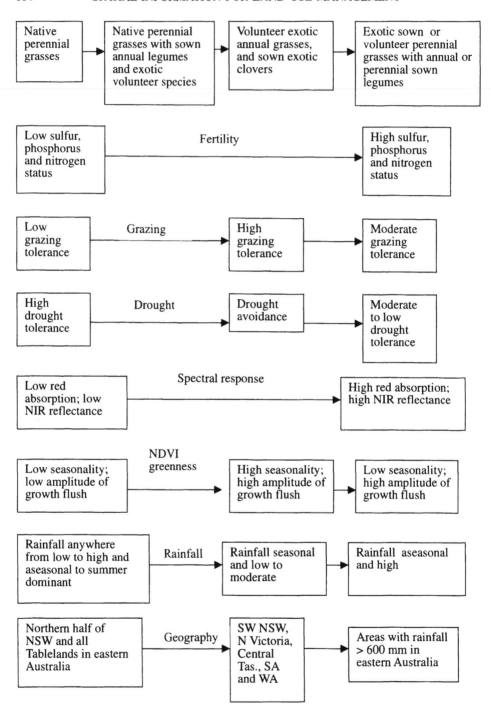

Figure 1. Interrelationships between environmental and management variables and characteristics of spatial data for Australian temperate pastures.

in species composition of pastures from total domination by native grasses to relative domination by exotic grasses, legumes and broadleaf volunteers (Figure 1). Associated with this change in species and fertility is an increase in absorption in the red and reflectance in the near infrared areas of the electromagnetic spectrum. As a result, a change in "greenness", as measured by a remotely sensed vegetation index, from low to high represents a change in species composition and growth potential as well as a change in green leaf area index. However, it could also represent the impact of seasonal rainfall on growth of uniform species in exotic pasture lands, or that of periodic rainfall on volunteer species inhabiting native perennial grasslands. This complex interrelationship between moisture, fertility and species provides both difficulties and opportunities for the use of satellite imagery to provide qualitative and quantitative estimates of grassland properties.

Framework for Approach

Information required for management decisions impacting pasture, grassland and the grazing industries may be divided into policy and operational areas. At the policy level, government agencies and commercial organizations may wish to know the land cover characteristics, the distribution of species, general productivity, species suitability, and extent of degradation with a view to sustainably managing land or selling seed, fertilizer and equipment. At the operational level, graziers may wish to know how much pasture is present and how fast it will grow; the location and size of problem areas (e.g. consistently poor pasture production due to edaphic factors); and what to do about these problem areas. The variation in the physical environment may be described by spatial data, however the effects on animal production are mediated through complex chemical, biological and behavioral processes which make linkages between spatially defined constraints and production or economic outcomes difficult. The process of developing spatial data applications for grassland monitoring and management is one of providing layers of evidence with specified confidence and error. In this context, there have been certain developments which have been of critical importance.

SOME CRITICAL TECHNOLOGIES FOR THE DEVELOPMENT OF GRASSLAND APPLICATIONS

Repeatable Classification of Grassland Condition with Remote Sensing

A foundation technology for spatial data applications in temperate Australian grassland was provided by classification of Landsat TM (Thematic Mapper) data to give a qualitative but objective and repeatable assessment of pasture growth potential which could be used as an indicator of superphospate fertilizer requirement (Vickery and Hedges, 1987; Vickery and Furnival, 1992). The classification was based on the transformation of the green, red and near-infrared channels of Landsat or SPOT satellite imagery by Principal Coordinate Analysis to create a two-dimensional data space based on the greenness and brightness (Hedges and Vickery, 1987; Figure 2). The resulting classification separates trees, woodland, water and sparse vegetation from a multi-class continuum ranging from the brownest to greenest pasture corresponding to the change from pure native to highly improved

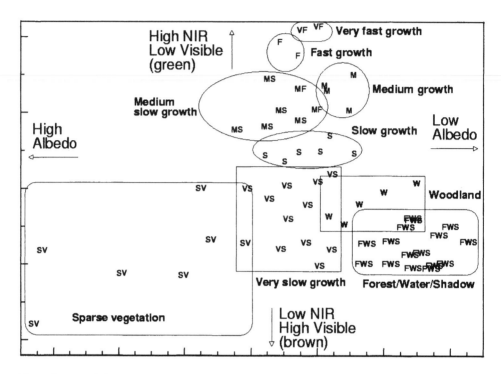

Figure 2.   Plot of class centroids on PCA plane showing aggregation into pasture growth status and other land cover groupings.

pasture described earlier (Plate 14 top left). This non-quantitative procedure has proven repeatable both temporally and geographically throughout the south east of Australia (Cayley *et al.*, 1989; Reid *et al.*, 1993). It has ultimately provided one of the few examples of a commercialized product for grazing agriculture derived from remote sensing.

Spatial Interpolation of Sparse Climate and Related Data

The development and implementation of spatial interpolation methods for generation of weather and climate layers using terrain features as covariates (Hutchinson, 1989) provides the means to describe the spatial properties of the climatic controls on grassland vegetation in data-sparse regions. This enabled the development of terrain-based climate thresholds where previously threshold boundaries were only a function of point distribution and a statistical algorithm. Such terrain-based climate thresholds provided the means for modelling potential suitability zones for pasture plants based on simple algorithms such as a growth index model (Nix, 1981; Hill *et al.*, 1996) or logical models (Hill, 1996). Subsequently, comprehensive Australia-wide coverage of near real-time climate data has enabled the implementation of modelling systems to predict drought and rangeland productivity at a continental scale (Hall *et al.*, 1997).

Time Series Analysis with NDVI

The Normalized Difference Vegetation Index (NDVI) is based on the difference in reflectance from green vegetation between the red and near infrared parts of the electromagnetic spectrum (NDVI = (NIR − Red)/(NIR + Red)). The NDVI increases with increasing green leaf area, but the relationship is curvilinear reaching an asymptote at a leaf area index of between 2 and 4. (see Friedl, 1997). Bimonthly composites of the NDVI at 1.1 km resolution have been constructed from NOAA AVHRR (National Oceanic and Atmospheric Administration Advanced Very High Resolution Radiometer) data for the whole of Western Australia (WA) since 1991. Such data are widely used in vegetation monitoring and have been the subject of substantial local research on rangeland and agricultural monitoring (Smith, 1994; Cridland et al., 1994; Roderick, 1994). The development of approaches to quantitative analysis of times series of NDVI imagery (Mallingreau, 1986; Reed et al., 1994) has enabled the extraction of mechanistically meaningful indices from these data, taking utility and interpretation beyond the qualitative greenness concept. Time series of AVHRR NDVI provide both spatial coverage and temporal pattern which may be interpreted to be associated with grassland with different growth patterns. The approximately 1 km resolution of these data means that grassland or pasture is mixed with other vegetation types within a pixel for all but the most extensive and uniform grassland areas eg. the USA Great Plains (Tieszen et al., 1997).

Probabilistic Approach to Heterogeneity and Uncertainty — Bayesian Inference

Bayes' theorem (Bayes, 1763 cited in Ellison, 1996) states that the joint probability of two events $P(\theta|x)$ equals the product of the probability of one of the events and the conditional probability of the second event given the first. Bayesian statistics provides a methodology for estimating the probability of presence of a feature based on a relationship between predictor data and data describing the feature. It provides one means of incorporating uncertainty into spatial representations of grassland. Since grassland are inherently variable and spatially heterogeneous, class boundaries relating to biomass, species composition or other properties become highly uncertain. If the presence or absence, or the quantitative value of a biophysical property can be expressed in terms of some probability threshold, the level of uncertainty is included in the spatial representation.

Bayesian inferencing systems are available in several GIS packages and usually require the estimation of joint or conditional probabilities of presence and absence, or of the occurrence of various classes or levels of a biophysical property. They have been used to analyse habitat relationships for red deer (Aspinall, 1992) and distribution of rangeland plants on the prairies of Alberta (Hill et al., 1997). A logistic response model has proven useful for developing joint probabilities of association between botanical composition and classified Landsat TM data (Vickery et al., 1997) for use in the inference of most probable pasture type.

Synthetic Aperture Radar

Synthetic aperture radar (SAR) is sensitive to the water content and structural characteristics of grassland vegetation and is unaffected by cloud. The wavelengths in the microwave part of the electromagnetic spectrum are between 1 and 100 cm, which almost exactly covers the range in height of all the world's grasslands from manicured lawns to tropical speargrass and temperate tussocks. The crucial feature of potential benefit from radar yet to be realized is the discriminatory ability of combinations of SAR frequencies and polarizations. At present these are only available on airborne systems and potential over grassland and pasture has been little explored. Over croplands, analysis of multi-frequency multi-polarized radar has developed to the level of attempting to model backscatter pattern based on structural differences between crops modelled as a series of cylinders and discs (*e.g.* Ferrazoli *et al.*, 1997). Heterogeneous grasslands may not lend themselves to this type of simulation. However, several factors suggest substantial potential for widespread use of a multi-frequency multi-polarized radar satellite:

1)   the relationship between the variation in thickness of grassland canopies, mostly between 1 and 50 cm, and the most common wavelengths of radar instruments of Ku band 1 cm, X band 3 cm, C band 5.6 cm, L band 23 cm likely to be combined on a satellite platform;
2)   the difference in response to horizontally and vertically polarized radar waves by horizontally and vertically oriented herbage;
3)   the impact of cloud as a major restriction on the use of optical data with grasslands and the difficulty of spectral reflectance modelling with grassland herbage.

CASE STUDIES INTEGRATING SOME OF THE CRITICAL TECHNOLOGIES

The case studies described cover a range of scales from continental to paddock and a range of clients from national committees to single grazing enterprises. However the case studies form a progression from: (1) a simple presence/absence type layer for plant distribution; to (2) a multi-class land cover classification; (3) a representation of a landscape-scale process in time; (4) a biophysical property such as biomass or height; (5) the output of a mixture of biological processes involved in grazing pasture; and finally (6) an analysis of pasture suitability based on a combination of the some of the previous case studies. The outcome is in the form of a map or maps which add increasing amounts of information to a decision process and become more prescriptive as the input information increases.

Mapping Potential Distribution of Pasture Species

Where knowledge of plant responses to climatic constraints are well known and broad coverage is required, simple logical models may be used to create maps of potential suitability zones for pasture plants (Hill, 1996). An example of the algorithm for alfalfa (*Medicago sativa*) in Australia is provided below together with the rationale for each part of the rules.

alfalfa = (summer P/E $\geq$ 0.2 and annual rainfall $\geq$ 300 mm)
or (summer P-E $\geq$ –80 mm and annual rainfall $\geq$ 300 mm) *summer dry but climate not too arid by two different measures*
or (summer P/E $\geq$ 0 and summer P/E < 0.5 and spring P/E $\geq$ 0.8 and autumn P/E $\geq$

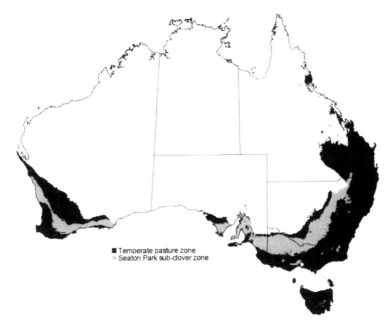

Figure 3. Map showing the potential distribution of Seaton Park subterranean clover in Australia relative to the whole temperate pasture zone.

0.8) *summer very dry but overall climate less arid*
and (winter P/E ≥ 0.3) *some winter moisture ie. Not into tropics but included dry subtropics*
and (annual rainfall < 1500 mm) *not extremely wet*

A zone resulting from a model such as this for the subterranean clover (*Trifolium subterraneum*) cultivar Seaton Park is shown in Figure 3 overlayed on the extent of the whole agricultural zone in southern Australia. Similar models have been used for native rangeland plants in Alberta (Hill *et al.*, 1997).

If data describing presence and/or abundance of species are available, then adaptation or suitability zones may be inferred from Bayesian inferencing by association of presence or abundance with other data layers describing climate or the physical environment (Hill *et al.*, 1997). Variables for zone prediction may be selected on the basis of the best Chi square scores for association with species sites (Aspinall, 1992). The resulting maps provide a graduated probability of location or suitability for plants, such as the distribution of Mixed and Fescue Prairie species in southern Alberta (Hill *et al.*, 2000).

Land Cover Classification of Grassland Types

Land cover classification at the continental scale provides the basic land area definition for higher level analysis. Meaningful class definition for cultivated pastures is difficult where the scale of land use variation is well below the resolution of imagery, such as NOAA

AVHRR NDVI, which provides the spatial and temporal coverage needed. If a robust, repeatable high resolution classification procedure such as the pasture growth status procedure described earlier (Vickery and Furnival, 1992) is available for pastures, it may be used to describe the proportions of grassland and other vegetative cover within low resolution AVHRR pixels. Cover classes may then be defined on the basis of most predominant grassland or other cover type as well as the temporal pattern of NDVI (Hill et al., 1999b). De-commercialisation of the Landsat program makes coverage with high resolution imagery affordable, however discrimination of grassland and pasture types is still enhanced by time series of imagery such as NOAA AVHRR which may detect differences in seasonal patterns. An example of a pastoral land cover class defined in these terms is given below (Hill et al., 1999b).

Class K	Improved Pasture
NDVI profile	Extended peak in spring and summer;
Within pixel TM classification characteristics	17% very fast and fast growth classes; 28% woodland and forest classes
Locational features	High elevation improved pastures associated with class L (a similar class with lower maximum NDVI).
Description	Improved perennial pastures with some woodland.

Regional Patterns of Pasture Growth

Environments where growth of grassland vegetation is seasonal are most suitable for quantitative analysis of NOAA AVHRR NDVI time series since calculation of curve features is not confused by multi-modal patterns. The south west of Western Australia with a climate characteristic of the Mediterranean region — highly seasonal, winter dominant rainfall with hot, dry summers — provides a good example. The association between the seasonal rainfall pattern and the high NDVI period in the center of the agricultural zone of Western Australia is shown in Figure 4 (top left). The time series is smoothed with a 5-interval running mean and forwards and backwards lagged moving averages are used to define the start and end of the growing season at the points where they intersect the smooth curve (Reed et al., 1994).

The lag period is usually set to an interval equivalent to the length of the non-growing season, a period largely defined by temperature in the continental U.S. (Reed et al., 1994). However, an environment such as the southwest of Western Australia where moisture availability defines the growing season presents additional difficulties. For example, the end of the growing season occurs when senescence of pasture begins to significantly exceed growth. However, in terms of NDVI curve features, it is represented by the time at which the NDVI reaches a value close to its minimum. This occurs when almost all trace of chlorophyll has gone from pasture, and long after actual growth potential has ceased. Additional climate variables such as a temperature-based heat sum and a moisture-

or phenology-based end of season date may be needed to provide seasonal limits for NDVI metrics (Hill *et al.*, 1998a).

In this environment, increments in NDVI between t_0 and t_1 should have some relationship to growth rate of annual pastures or grassland, but the scale of available time series data from NOAA AVHRR presents significant problems since 1 km pixels are usually a mixture of pasture, cropland, trees or remnant vegetation. We have found that in spite of scale problems, field growth rate estimates can be related to NOAA AVHRR NDVI, with additional variation explained by differences in bi-weekly mean air temperature and changes in phenology of plants between winter and spring (Figure 4 top right; Donald, unpublished data). In winter, NDVI tends to increase with little increase in pasture growth rate due to temperature constraints. In spring, growth rates increase due to reproductive development, but there is little increase in NDVI since leaf production has largely ceased.

Estimation of Biophysical Properties of Pasture

Estimation from spectral properties

Measurement of biophysical properties of grassland with remote sensing can provide direct inputs into management decision making. As early as 1973, Pearson and Miller working with the International Biological Program (IBP) described a hand held "biometer" for estimation of grassland biomass using the ratio of the difference in reflected red and NIR radiation; a vegetation index like the NDVI. Whilst quantitative estimation using the NDVI has not generally reflected the original promise, in certain circumstances it can be a reliable biomass estimator.

Basic feed budgeting to optimize the pasture presented to sheep requires information on how much feed is present at time t_0 and what the growth rate will be between t_0 and t_1. The NDVI is related to the LAI (leaf area index) of the pasture sward and therefore also to the standing biomass, whilst the pasture is still actively growing. For annual pastures in highly seasonal environment such as the agricultural zone of WA (see Figure 3), the pasture behaves much like a crop and NDVI is well correlated with biomass up until the peak of the growing season. A useful predictive relationship may be derived (Figure 4 bottom left) provided; (1) NDVI is only required to estimate membership of a pasture class having a biomass within some specified range, such as 200 kg/ha intervals; and (2) the pixel estimates are integrated to provide a single paddock value (Hill *et al.*, 1998b). Examples of equations relating feed-on-offer (FOO kg dry matter/ha) to NDVI calculated from Landsat TM imagery in spring of 1995 near Mt Barker, Western Australia are shown the table below.

Image Date	FOO sampling date	Equation range (kg/ha)	FOO interval (kg/ha)	Equation	R^2
August 27	August 14	200–3000	200	FOO = −16205 + 23626 (NDVI)	0.76
October 14	October 3	1500–3500	250	FOO = −14538 + 24108 (NDVI)	0.90

This type of relationship only applies to green, grazed pasture; once stem elongation and flowering commence, or soil moisture declines, or if pasture is ungrazed and accu-

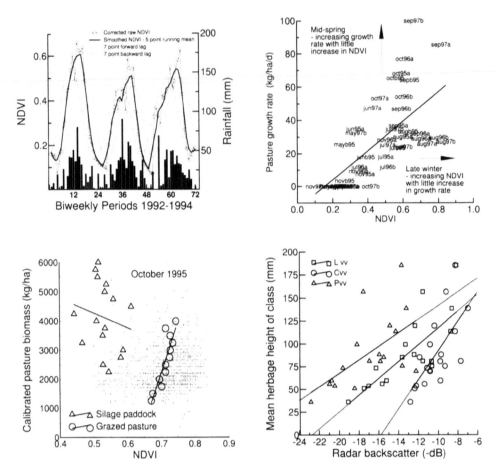

Figure 4. Bottom right. Relationship between herbage height and radar backscatter for perennial grassland on "Chiswick" near Armidale, NSW. Bottom left. Relationship between NDVI and feed-on-offer for August and October 1995 for annual pasture at Mt Barker, WA based on grouping ground data into 200 kg/ha classes. Top left. NDVI profiles for 1992 to 1994 for a point near Katanning, WA showing the corrected raw NDVI, smoothed NDVI, and forward and backward lag curves used to define the beginning and end of the NDVI "green period". The bars indicate bi-monthly rainfall for the same period showing the seasonal climate. Top right. Relationship between pasture growth rate and NDVI from NOAA AVHRR bi-weekly images for three properties in the agricultural zone of Western Australia.

mulated for silage making (Figure 4 bottom left), the relationship does not apply. One key to practical use is to generalize the change in slope and offset for such curves relative to the date in a standardized season. This can be done in Western Australia, however several issues remain as obstacles to practical application: (1) the difficulty of establishing an offset value using limited ground data; (2) the decline in sensitivity to biomass of NDVI as the season progresses; (3) the difficulty in acquiring imagery at the start of a season due to cloud cover; and (4) the difficulty in interpolating FOO estimates between infrequent satellite dates due to the temporal variability in the spatial pattern of pasture response induced by grazing.

Estimation from structural properties

Optical data have limitations when dealing with high biomass loads, dry or senescent material, mixtures of green and dead or measurement in cloudy environments. However synthetic aperture radar is not affected by cloud. An instrument with multiple frequencies and polarizations, such as the NASA/JPL (National Aeronautic and Space Administration Jet Propulsion Laboratory) airborne SAR system is capable of providing detailed quantitative information on grassland vegetation. Combination of backscatter returns from C (6 cm) L (23 cm) and P (68 cm) bands with both vertical and horizontal polarization can provide a high degree of discrimination between grassland types when these types are distinguished by clear differences in canopy structure (Hill *et al.*, 1999a). Radar backscatter is well correlated with herbage height of temperate pastures (Hill *et al.*, 1999a; Figure 4 bottom right) and may be used to map height where grasslands are diverse and variable (Plate 14 bottom left). This sensitivity to structure and height suggests that biomass will also be retrievable. Current satellites with single frequency C band systems are not useable for this purpose, but new instruments such as ENVISAT with multiple frequencies or polarizations should make such measurements routinely available.

Coupling Spatial Data with a Grazing System Model

The outputs and utility of simulation models of animal production and pasture growth may be enhanced by coupling with spatial data. This is particularly so where soil or terrain maps can provide spatial distribution of quantitative estimates of properties which may directly influence the parameters in the model. Ideally, the model should be designed with a spatial context in mind. It may operate more easily at a landscape scale with well defined physical constraints (*e.g.* Namken and Stuth, 1997). Coupling of spatial data with agricultural models at field scale has been demonstrated recently with polygons of soil attributes and a typical, detailed, mechanistic, crop simulation model for peanuts (Engel *et al.*, 1997). However, spatial representation of vegetation properties is harder to obtain since remote sensing has had little success in retrieving quantitative estimates of the parameters and variables which are actually used in mechanistic simulation models.

It is possible to use paddock maps of pasture type and a pasture growth status classification of Landsat TM imagery (Plate 14 top left) to provide a qualitative representation of spatial variation in grassland. As previously explained, fertility classes may be derived, and pasture condition (Plate 14 top right) may be inferred from the pasture growth status classification (Vickery *et al.*, 1997; Hill *et al.*, 1999c). The spatial data inputs and outputs are loosely coupled to the model by tables of input parameters and look-up tables of output values which may be displayed by reclassification of a template map as shown in the table below.

Process output	Basic and derived	Model input map input	Model output	Derived map
Survey	# Paddock map of pasture type	# Plant parameters	Look-up table of simulated biomass	E.g. Maps of NPP for selected
Remote sensing	# Pasture growth map	# Fertility index	aggregated over time intervals	time intervals using combined
Bayesian inference	# Most probable pasture condition	# Pasture composition	for pasture type and pasture class	map inputs as template

The results might be a representation of annual net primary production (Plate 14 bottom right) whose spatial veracity is dependent of the accuracy of the input maps and the accuracy of the simulation model. The weakness in this approach is that neither spatial data handling nor the simulation model has been designed with the other is mind. This issue has been dealt with in detail by Bennett *et al.* in Chapter 6.

Analysis of the Pastoral Potential of a Large Property

Spatial information often forms the basis for property planning. However, this is frequently based on descriptive boundary, terrain and land cover data alone. Planning may be enhanced by making use of spatial data which represent the outcome of a biological process which occurs through time (*e.g.* the output of a growth model). At "Cooplacurripa", on the east coast of New South Wales, remote sensing, terrain data, climate data, simple modelling, and GIS procedures were combined to provide an assessment of the suitability of forage plants for the property (Hill *et al.*, 1996). This assessment of agricultural potential was then used to define areas for clearing, subject to a decision procedure, including land capability and fauna impact assessment which integrates agricultural and environmental priorities.

Cooplacurripa is located in the edge of the escarpment of the Great Dividing Range in eastern Australia. This provides an unusual physical environment for a single property since elevation varies between about 200 and 900 m over the whole area. Assessments were made in two main areas:

1) the condition of the existing pasture; and
2) the suitability of the environment for growth of the main groups of pasture plants as defined by thermal response.

Using the classification procedure for Landsat TM imagery described previously, tree cover and pasture area were mapped to show the pasture growth status of the current pasture (Plate 15 top left). Attention for the analysis was focused on the northern part of the property where rainfall was highest and average annual temperatures were lowest (Plate 15 top right). A pasture suitability map calculated using climate surfaces, a growth index model (Nix *et al.*, 1989), and estimates of potential growth rates for typical species defined much of this area as highly suitable for cool season pasture plants (Plate 15 bottom left). Much of the existing cleared area was best suited to warm season pasture plants, but a significant amount of the treed area was suitable for cool season species even when a slope constraint was applied (Plate 15 bottom right).

A consulting report was commissioned by the manager of Cooplacurripa to assess the environmental constraints to clearing or thinning trees in this part of the property. When the guidelines for exclusion of clearing where constraints exist (erosion hazard, chemical or physical limitations to pasture, mass movement hazard, slope exceeds 18°, sedimentation and stream band erosion hazard, landscape amenity compromised, rockiness limitation and fauna habitats) were applied, the proportion of the study area available for clearing was reduced to only 5% or about 220 ha scattered over a wide area (Andrews *et al.*, personal communication). The table below shows the contrast between the potential pasture area and the area clearable within land and fauna constraints.

Total (ha)	Forest (%)	Grassland or Open woodland (%)	Treed area < 20 deg slope suitable for pasture (%)	Area suitable for clearing within land and fauna constraints (%)
3715	63	37	29	5

The study provided a good example of the power of combining spatial analysis from two perspectives, production and conservation, to obtain a socially and environmentally acceptable outcome.

SOME LESSONS FOR SPATIAL DATA HANDLING FOR GRASSLANDS

The case studies highlight some important issues which require consideration in the application of satellite-derived spatial data for grassland management.

1) The sensitivity and precision of the detection system. *e.g.* NDVI is saturated at high leaf areas and short wavelength SAR is completely attenuated in the canopy at relatively low herbage thickness or volume.
2) The spatial scale of variation detected by the sensor and actual variation in meaningful biophysical properties of grassland. *e.g.* NDVI is insensitive to wide variation in biomass of senescent vegetation.
3) The importance of texture measurements and texture operators in retrieving information from imagery over grassland. *e.g.* Speckle reduction and textural filtering are needed to generalize SAR data over grassland.
4) The spatial scale of differences in biophysical properties needed for practical/commercial significance. *e.g.* Clear differences in spectral response between patches may or may not correspond with clear differences in animal performance.
5) The importance of appropriate temporal frequency for capture of the dynamics of the herbage growth/animal grazing process. *e.g.* Clear difference in annual pasture biomass between paddock units based on topography and soils are not reflected in single date NDVI images.
6) The availability of the appropriate ground truth measurements to match the sensor characteristics. *e. g.* Structural data such as leaf angle distribution, bulk density and herbage moisture profiles are rare for grasslands but important for optical and SAR image calibration.

CONCLUSIONS

This chapter has described development of various spatial information layers which may be used as evidence for management decisions at a range of scales for the grazing industry in temperate areas of Australia. Some aspects are unique to the Australian context, but others have universal application. The information layers developed owe more to the implementation of established remote sensing and spatial data handling methods than they do to sophisticated biophysical modelling of reflectance or complex spatial analysis. Many involved implementation of very simple, well established modelling and processing approaches. Whilst more sophisticated methods may be available, adapting these to the scale, heterogeneity and spatial variability of grassland often constitutes a difficult challenge. The development of more advanced satellite sensors with hyperspectral coverage, hyperspatial resolution, very high frequency, fully polarized and multi-frequency radar, and laser systems for direct measurement of canopy structure and height seems to promise much for the future. However, the variability in chemical constituents, species composition and physical structure of grassland and pasture will continue to present a significant challenge.

ACKNOWLEDGEMENTS

The work described here is the result of many years of effort by colleagues and collaborators in CSIRO Animal Production. I wish to thank Peter J. Vickery, Graham E. Donald, E. Peter Furnival, Colin Mulcahy, Asoka Edirisinghe. Development of the pasture classification was supported by the WoolMark Company; pastoral land cover mapping was supported by Meat and Livestock Australia; mapping pasture species distribution was supported by the Meat and Livestock Australia, The Grains Research and Development Corporation, the Dairy Research and Development Corporation and the Woolmark Company. Biomass and growth rate estimation in Western Australia is carried out in collaboration with Agriculture WA and the WA Department of Land Administration. CSIRO Animal Production is partially funded by Australian wool growers through the Woolmark Company.

REFERENCES

1.   R.J. Aspinall (1992) *Int. J. Geog. Inf. Syst.*, **6**, 105–21.
2.   J.W.D. Cayley, P.J. Vickery and D.A. Hedges (1989) in *Proceedings of the 5th Australasian Remote Sensing Conference, Perth.* (Western Australia: Committee of the 5th Australasian Remote Sensing Conference), p. 534.
3.   S.W. Cridland, D.G. Burnside and R.C.G. Smith (1994) in *Proceedings of the 7th Australasian Remote Sensing Conference*, Melbourne, Australia. (Western Australia: Remote Sensing and Photogrammetry Association of Australia), pp. 1134–1141.
4.   T. Engel, G. Hoogenboom, J.W. Jones and P.W. Wilkens (1997) *Agron. J.*, **89**, 919–28.
5.   A.M. Ellison (1996) *Ecol. Appl.*, **6**, 1036–1046.
6.   R.D. Graetz, R.P. Pech, N.L. Hindley and M.R. Gentle (1984) The application of Landsat to rangeland management: Libris five years on, in *Proceedings of the 3rd Australasian Remote Sensing Conference, Gold Coast, Queensland*, edited by E. Walker, pp. 322–331. (Brisbane, Australia: Organizing Committee of LANDSAT 84).
7.   P. Ferrazzoli, L. Guerriero and G. Schiavon (1997) A vegetation classification scheme validated by model simulations, in *IGARSS'97, Proceedings of the 1997 International Geoscience and Remote Sensing Symposium*, Singapore, edited by T. L. Stein, Vol 4, pp. 1618–1620. (Texas: IEEE Geoscience and Remote Sensing Society).
8.   M.A. Friedl (1997) Examining the effects of sensor resolution and sub-pixel heterogeneity on spectral vegetation indices, in *Scale in Remote Sensing and GIS*, edited by D.A. Quattrochi and M.F. Goodchild, p. 129. (Boca Raton: CRC Lewis Publishers).

9. W. Hall, K. Day, J. Carter, C. Paull and D. Bruget (1997), Assessment of Australia's grasslands and rangelands by spatial simulation, in *MODSIM '97, International Congress on Modelling and Simulation, Proceedings*, edited by D. McDonald and M. McAleer, Vol. 4, pp. 1736–1741. (University of Tasmania: Modelling and Simulation Society of Australia).

10. D.A. Hedges and P.J. Vickery (1987) Use of a principal components strategy as the basis for an unsupervised classification routine to examine Landsat data from grassland vegetation, in *Proceedings of the 4th Australasian Remote Sensing Conference, Adelaide*, edited by D. Bruce, pp 178–188. (Adelaide: 4th Australasian Remote Sensing Conference Pty Ltd).

11. M.J. Hill (1996) *Aust. J. Agric. Res.*, **47**, 1095–1117.

12. M.J. Hill, G.E. Donald, P.J. Vickery and E.P. Furnival (1996) *Aust. J. Exp. Agric.* **36**, 309–321.

13. M.J. Hill, R.J. Aspinall and W.D. Willms (1997) *Ecol. Model.*, **103**, 135–150.

14. M.J. Hill, G.D. Donald and R.C.G. Smith (1998a) NDVI-based phenological indices: predictive potential for grazing systems, in *Proceedings of the 9th Australasian Remote Sensing and Photogrammetry Conference*, CDROM. (Sydney: Remote Sensing and Photogrammetry Association of Australasia).

15. M.J. Hill, G.E. Donald, G.A. Wheaton, M. Hyder and R.C.G. Smith (1998b) Remote sensing for precision pasture management in south western Australia, in *Proceedings of the 9th Australasian Remote Sensing and Photogrammetry Conference*, CDROM. (Sydney: Remote Sensing and Photogrammetry Association of Australasia).

16. M.J. Hill, P.J. Vickery, E.P. Furnival and G.E. Donald (1999a) *Remote Sens. Environ.*, **67**, 15–31.

17. M.J. Hill, P.J. Vickery, E.P. Furnival and G.E. Donald (1999b) *Remote Sens. Environ.*, **67**, 32–50.

18. M.J. Hill, G.E. Donald, P.J. Vickery, J.R. Donnelly and A.D. Moore (1999c) *Aust. J. Exp. Agric.*, **3**, 285–300.

19. M.J. Hill, W.D. Willms and R.J. Aspinall (2000) *Plant Ecol.*, **127**, 59–76.

20. M.F. Hutchinson (1989) *CSIRO, Australia, Division of Water Resources Technical Memorandum*, **89/5**, 95–104.

21. J.P. Mallingreau (1986) *Int. J. Remote Sens.*, **7**, 1121–1146.

22. J.C. Namken and J.W. Stuth (1997) *Int. J. Geog. Inf. Sci.*, **11**, 785–798.

23. H. Nix (1981) Simplified simulation models based on specified minimum data sets: the CROPEVAL concept, in *Application of Remote Sensing to Agricultural Production Forecasting*, edited by A Berg, pp. 151–169. (Rotterdam: A.A. Balkema).

24. R.L. Pearson and L.D. Miller (1973) Colorado State University, Dept. Watershed Sci, Incidental Report no. 6.

25. B.C. Reed, J.F. Brown, D. VanderZee, T.R. Loveland, J.W. Merchant and D.O. Ohlen (1994) *J. Veget. Sci.*, **5**, 703–714.

26. R.N.D. Reid, P.J. Vickery, D.A. Hedges and P.M. Williams (1993) *Aust. J. Exp. Agric.*, **33**, 597–600.

27. M.L. Roderick (1994) PhD Thesis, School of Surveying and Land Information, Curtin University of Technology, 349 pp.

28. R.C.G. Smith (1994) *Australian Vegetation Watch, Vegetation Watch*, Final Report RIRDC reference No. DOL-1A.

29. L.L. Tieszen, B.C. Reed, N.B. Bliss, B.K. Wylie and D.D. DeJong (1997) *Ecol. Appl.*, **7**, 59–78.

30. P.J. Vickery and D.A. Hedges (1987) Use of Landsat MSS data to determine the fertilizer status of improved grasslands, in *Proceedings of the 4th Australasian Remote Sensing Conference, Adelaide*, edited by D. Bruce, pp. 287–96. (Adelaide: 4th Australasian Remote Sensing Conference Pty Ltd).

31. P.J. Vickery, D.A. Hedges and M.J. Duggan (1980) *Remote Sens. Environ.*, **9**, 131–48.
32. P.J. Vickery and E.P. Furnival (1992) Development and commercial use of Landsat derived maps as an aid to more effective use of fertilizer, in *Proceedings of the 6th Australian Society of Agronomy Conference*, pp. 392–395. (Armidale: Australian Society of Agronomy).
33. P.J. Vickery, M.J. Hill and G.E. Donald (1997) *Aust. J. Exp. Agric.*, **37**, 547–562.

10. AN INTELLIGENT SYSTEM FOR MONITORING FORESTS

David G. Goodenough, A. (Pal) S. Bhogal, Daniel Charlebois, Andrew Dyk and Richard J. Aspinall

INTRODUCTION

Sustainable development is defined as a process of using resources, such as forests, to meet the needs of the present generation without compromising the ability of future generations to meet their needs (Canadian Council of Forest Ministers 1995, WCED 1987). Sustainable development remains one of the key issues for the new millennium. Sustainable use of forests is particularly important to Canada. About 418 million hectares are forested in Canada; this is about half the country and 10% of the world's forests. Canada is the worlds largest trader of forest products, with over 20% of the world trade. The forest sector contributes more than $70 billion to Canada's GDP; forests contribute more to Canada's balance of trade than agriculture, fisheries, mining and energy combined. Approximately 850 000 Canadians are employed in forestry and related industries, and almost 350 communities are forestry-dependent (Goodenough et al., 1999).

Canada's commitment to sustainable development has been demonstrated through a number of initiatives including a national forest strategy in 1992, ratification of the Convention on Biological Diversity in 1994, preparation of the Canadian Biodiversity Strategy in 1995, and participation in the Montreal Process on Criteria and Indicators. Under the 1997 Kyoto Protocol, Canada agreed to reduce greenhouse gas emissions by 6% of 1990 levels within 11 years (Goodenough et al., 1999). Nations are expected to modify their national forest inventories to include measurements to meet the reporting needs of the Protocol. The Criteria and Indicators initiative, for which the Canadian Forest Service provides leadership, was established by the Canadian Council of Forest Ministers (1995) and requires periodic reports to the international community on the condition of our forests. These reports must be based on accurately defined measures. The reporting aspects of the Kyoto Protocol and other initiatives ensure that inventory and reporting have become important aspects of Canada's agenda for forestry.

There are three separate reporting requirements in the Kyoto Protocol as it relates to forests: 1) reforestation, afforestation, and deforestation; 2) carbon stocks since 1990; and 3) land use change and forest inventory. Concern for sustainability has also led to studies aimed at improved understanding of sustainable forest management and its impact on Canadian resources and the environment.

As the second largest country in the world and with most of the territory being very sparsely populated, Canadian governments need to rely on satellite observations as a principal data acquisition tool. Earth observations and associated technology are therefore essential to provide:

1) timely and consistent information for national, regional or provincial planning and strategic decision making;
2) information on the current state of the Canadian forests and environment in response to the national and international reporting requirements;
3) data for resource monitoring, assessing the magnitude and impact of environmental change, impact assessment, and research; and
4) data for developing and delivering the government's environmental information and prediction strategies.

Monitoring the environment of a substantial geographic area, such as a province, nation, continent, or the globe requires the integration of data, information, knowledge, and expertise from many sources. The process of integrating data from multiple sources is data fusion. It is necessary since single data sources may not capture all the significant characteristics needed to identify an object and be subject to systematic errors. Multiple sources provide complementary information, and built-in redundancy increases the accuracy and reliability of interpretation. Historical data are also needed for detecting change over time. Data fusion is however, a complex problem. A number of computing and other scientific tools are relevant to data integration including artificial intelligence (planning, case-based reasoning, software agents, machine learning), remote sensing, geographic information science (UCGIS, 1996) and database management systems. As an example of the scale of the problem, the forest inventory of British Columbia contains around 7 000 1:20 000 maps or GIS files. These data must be maintained and updated to provide the information needed under various reporting and inventory schemes. The remainder of this chapter describes a system for intelligent data fusion in the context of forestry. The system draws on a full range of advanced techniques in computer science.

SYSTEM OF EXPERTS FOR INTELLIGENT DATA MANAGEMENT: SEIDAM

In 1990, the Canada Centre for Remote Sensing (CCRS) proposed the System of Experts for Intelligent Data Management (SEIDAM) project to NASA's Applied Information Systems Research Program. Seven partners came together to create and execute this project: CCRS of Natural Resources Canada; the Pacific Forestry Centre of the Canadian Forest Service; Industry, Science and Technology Canada; the British Columbia Ministry of Forests; the British Columbia Ministry of Environment, Lands and Parks; the European Union Joint Research Centre at Ispra, Italy; and the Royal Institute of Technology in Stockholm, Sweden.

SEIDAM addresses several research and operational problems, including fundamental issues of intelligent data management, and the processing and fusing of multi-temporal forest inventory and remote sensing information. As examples of some of the research issues, we highlight machine learning, spatial knowledge acquisition, use of digital terrain models, and image segmentation and analysis. SEIDAM specifically addresses data management (55%), visualization (25%), distributed applications (15%), and parallel processing (5%) in the context of forest and environmental monitoring.

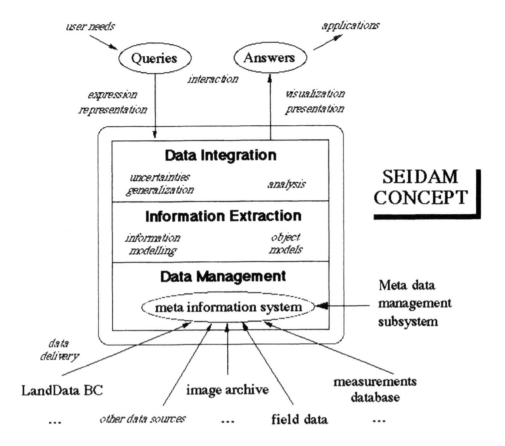

Figure 1. Conceptual model of SEIDAM. The model shows the integration of GIS, remote sensing, modeling, and database and visualization technologies in an artificial intelligence environment created using PROLOG. Diverse data sources are managed (using a metadata management system) and analyzed by SEIDAM in response to user queries and product requests.

SEIDAM Structure

Figure 1 shows the design structure and Figure 2 (top) the architecture of SEIDAM. Raw and processed forest inventory and remote sensing data, as well as metadata describing the data sources, are stored in database repositories external to SEIDAM. Both the image and GIS metadata comply with Federal Geographic Data Committee (FGDC) standards (FGDC, 1997). The image metadata are created using extensions to the FGDC standard. SEIDAM's Metadata Management System (SMMS) allows metadata to be automatically acquired on data ingest from a variety of media and links the infrastructure components (metadata, image, and image overview) together. The SMMS also facilitates data availability on demand as SEIDAM automatically seeks new images to create a new product.

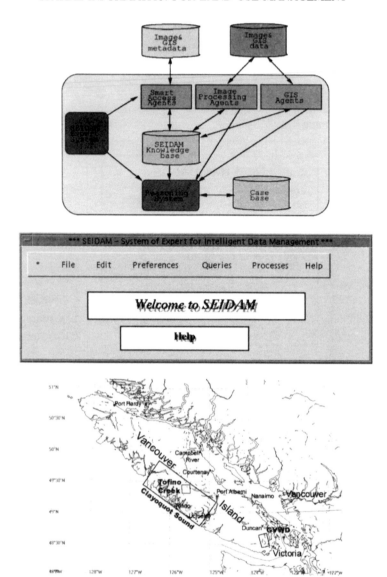

Figure 2. Top. SEIDAM system architecture and components. The top level SEIDAM expert system organizes from a collection of expert systems (agents) a plan to create the desired product or answer to a query. The plan is created by the reasoning system drawing on past training and experience (case base) and knowledge. Each agent can control one or more programs, such as GIS and image analysis, to accomplish a task. Middle. Menu interface of the main SEIDAM window. SEIDAM interfaces are created using PROLOG in a UNIX environment. As users train the system, appropriate functions are enabled in the main interface window. Functions in this interface window may also be expanded to include applications other than forestry. Additional examples of windows can be found for SEIDAM at http://www.aft.pfc.forestry.ca. Bottom. Principal SEIDAM test sites on Vancouver Island, BC, Canada. The Greater Victoria Watershed District (GVWD) test site has historical GIS and forest mensurational data spanning forty years. The Clayoquot Sound and Tofino Creek test sites are representative of the mid-latitude West Coast rain forests of North America. Tofino Creek is an area for which extensive experiments on data fusion, hyperspectral analysis, and AirSAR analysis have been conducted.

SEIDAM contains seven individual components (Figure 2 top): the smart access software agents, the SEIDAM expert system, the reasoning system, the image processing software agents, the GIS software agents, the SEIDAM knowledge base and the case-based reasoning system. A complete description of these components and the reasoning system can be found in (Charlebois *et al.*, 1996a, b).

Management of SEIDAM components is by an expert system shell (RESHELL) written in Prolog. The first hierarchical expert system created was called the Analyst Advisor (Goldberg *et al.*, 1985; Goodenough *et al.*, 1987). This was a system intended to assist users with analysis of Thematic Mapper imagery and GIS. Subsequently, a second hierarchical expert system, System of Hierarchical Experts for Resource Inventories (SHERI), an intelligent system for forest inventory updating, was developed. SHERI was an important predecessor of SEIDAM. Both SHERI and SEIDAM use RESHELL to instantiate expert systems.

Case-based Reasoning

An important component of SEIDAM is a case-based reasoning system called PALERMO (Planning And LEarning for Resource Management and Organization) (Charlebois *et al.*, 1996a). This has recently been modified to include a non-linear total order planner (Minton *et al.*, 1994). The new system, PALERMO, uses a learning component based on explanation-based generalization that stores generalized solutions created by the planner. Solution retrieval and re-use are integrated with the reasoning system allowing SEIDAM to make search decisions based on past experiences (Goodenough *et al.*, 1998).

SEIDAM Use

SEIDAM begins with the user specifying their goals in the form of queries for the objects of interest, or by selecting products. Goals are used by the case-based reasoning component of SEIDAM, PALERMO to derive the best plan and organize which experts to invoke for data acquisition, data preparation, analysis, and output. These expert systems may invoke other expert systems, or invoke experts that control existing image analysis processes, geographic information system processes, models, and field measurement analysis processes. The Smart Access software agents interact with the user to convert the user's goal statement into queries applied to the image and GIS metadata databases. The objective of these queries is to extract all of the information contained in the databases that is relevant to the user's goal. This information is placed in the SEIDAM knowledge base and used by the reasoning system as its initial world. Software agents used by SEIDAM were specifically designed to work in conjunction with the problem solver. Each software agent is composed of an expert system, a knowledge-base, a Prolog module containing a set of operators and a set of programs (*e.g.* image analysis or GIS software). All of the agents were created with our training system called PAROT. This ensures that all agents are consistent with the requirements of PALERMO. The expert system component of the agent uses a set of procedural rules or sequences of steps that represent states and state transitions. A procedural rule normally fires an operator and then sets the next state according to the reaction of the program (a transition). This behavior is a finite state machine (FSM) where each transition is deterministic.

In a typical scenario for SEIDAM, a user selects a desired product that may be used to make land use decisions. The product is selected via a graphical user interface. An example of a product is a digital forest cover map updated with TM imagery in order to determine change in the total area of forest. The creation of this product becomes the goal that SEIDAM submits to the reasoning system. SEIDAM's main expert system activates the smart access software agent to retrieve relevant metadata from the remote sensing image and GIS metadata databases and stores the information in SEIDAM's knowledge-base. The goal statement describing the product is then submitted to the reasoning system and the reasoning system creates a plan to satisfy the user's goal. If the reasoning system is successful, the plan is executed. The execution of the plan entails activating each agent in the order prescribed by the ordering constraints included in the plan. As the agents execute, they access, process and create new image and GIS data. After the successful execution of all of the agents, a product satisfying the user's goal is created, and there is new information contained in the knowledge-base that must be stored in the metadata databases. SEIDAM will therefore activate the smart access agent to update the metadata databases.

Data Types

Four main types of data are used by SEIDAM: satellite data, airborne data, GIS data files, and field data. Sources of imagery are shown below.

Satellite data	Landsat Multi-Spectral Scanner and Thematic Mapper; Systeme Pour l'Observation de la Terre SPOT 3; European Space Agency ERS-1 and ERS-2 C band radar; Japanese Space Agency JERS-1 L band radar; and U.S. National Oceanographic and Atmospheric Administration Advanced Very High Resolution Radiometer.
Airborne data	CCRS synthetic aperture radar (X, C – dual polarization); CCRS MEIS (stereo push-broom scanner); CCRS AMSS (11-channel scanner); NASA/JPL airborne synthetic aperture radar (C, L, P-bands, fully polarimetric); NASA/JPL Advanced Visible InfraRed Imaging Spectrometer; NASA MODIS airborne simulator; NASA Airborne Ocean Color Imager; Borstad Compact Airborne Spectral Imager; and Oregon State University Ultra-Lite.

Management and Integration of Multi-temporal and Multi-source Data

The objective of data integration is to bring together multi-temporal data from diverse sources for optimizing information content. Earth science applications use data from multiple sources to improve the accuracy of object recognition, obtain historical geographic data for detection of change over time, and to distribute derived information. To integrate data properly, the following capabilities are required:

1) identification of an optimal subset of the available data and sources for integration;
2) estimation of the levels of noise and distortions due to sensor, processing, and environmental conditions when the data are collected;
3) selection based on spatial resolution, spectral resolution, and accuracies and limitations of the data;
4) retrieval and selection independent of data formats, archive systems, storage media and dataset location;
5) optimization of the computational efficiency of the integrated data sets to achieve the goals of the users.

The identification of an optimal subset of data sources for integration has traditionally been done manually. An analyst selects the remote sensing images and ancillary data for the analysis. The selection criteria depend upon a number of human factors. Such a method is effective if the number of data sources is small. In this case the number of factors to be considered are relatively manageable and the performance of the analyst is consistent and unbiased. Where there are analysts of different experience, and the number of competing data sources is large (coupled with requirements which are dynamic), then an expert system can help in the selection of the optimal subset. The criteria used for dataset selection include the detectability of the required objects, the ability to identify (separate) objects of interest, the accuracy of object recognition taking into account data quality, the expertise and accuracy requirements of the users, the cost of information extraction from multiple data sources and the effectiveness of the analysis.

In SEIDAM the data selection process is performed in two stages. The first stage determines the suitability of individual data sources to detect the objects of interest. For example, if a user wishes to use space-based remote sensing to detect wetlands in western North America, SEIDAM would choose sensors having spatial characteristics applicable to the detection of wetlands, spectral responses appropriate for detecting standing vegetation, and radar data obtained at wavelengths long enough to penetrate vegetation and to be reflected from standing water. Data sources which initially pass the selection criteria are subjected to the next selection stage where incompatibilities and costs of integration, are evaluated. An optimal subset of the data is selected for integration using case-based reasoning. SEIDAM uses past experience (cases) and constructs an analysis plan to achieve the user's goals.

SEIDAM handles data integration on an "as-needed" basis and is able to integrate: (1) point, line, and symbolic data from a GIS; (2) multiple sets of digital images collected at different times with different platforms; (3) digital elevation models; (4) climatic and soil information. SEIDAM also integrates diverse knowledge sources through its expert systems, each of which contains distributed knowledge. SEIDAM only loads expert systems if the user's goals and analysis plan require it. If, during the course of the analysis, SEIDAM is unable to achieve the accuracy goals set by the user, then the system will backtrack to select other data sources and tasks and create a new plan to recognize the objects sought.

Cartographic data are traditionally represented as points, lines and regions. SEIDAM selects and executes existing GIS to convert mapping projections, create specific digital overlays, load attributes into relational databases, and convert symbolic features into raster representations according to feature descriptors. Whenever necessary, SEIDAM calls existing software routines to perform preprocessing for imagery data. These include corrections for sensor noise (such as drop outs, stripping, dark current), geometric corrections for distortion due to earth curvature, platform attitude, and topographic effects, atmospheric corrections where calibrations are needed for temporal analysis; and radiometric corrections for illumination effects and antenna attenuation in the case of microwave sensors. We have previously developed expert systems for detecting clouds, cloud shadows, snow, smoke, and haze in Thematic Mapper imagery. This detection allows us to dynamically restrict analysis to areas where the data quality is best, and to build an agenda for future data source selection should appropriate data, free of clouds over the areas of interest, became available. SEIDAM is currently restricted to working on land information corresponding to a single map at a time, in order to ensure that the output GIS files have polygon information corresponding to a single date. Digital terrain models are used to ensure correspondence of data affected by topography. This includes estimation of shadows, positional shifts, and correction of distortions of physical dimensions and shapes.

Data integration can propagate data errors and uncertainties into the final product. When possible, data that best meet the specific accuracy goals are used in the data analysis process. SEIDAM can make use of certainty methods and conflict resolution amongst competing hypotheses (resulting from parallel execution of several expert systems) to maintain as high an accuracy as possible. This problem remains an important research area. Machine learning methods help focus on the optimal path for data integration.

Within SEIDAM, users specify their desired goals either by query from a list of queries or by specifying a product goal. In the case of forestry, a product goal of creating a current forest description would involve sub-goals such as the identification of the GIS file(s) or map(s) to be updated, the surface classes to be recognized, the level of accuracies desired, and the output visualization wanted. Given these goals, SEIDAM queries the directories of remote sensing data to identify potential imagery needed. If the imagery is available, SEIDAM identifies where the imagery is and transfers it across a network. SEIDAM can handle various computer storage formats. A geographic window can be defined such that only data falling inside the window are extracted. Future improvements to SEIDAM are to access data from distributed clearinghouses and to distribute agents through the Internet.

Data Volume and Storage

The amount of data to be collected in the 21st Century will exceed 1 terabyte of data per day. This growth in data volume will occur with the launch of the EOS satellites and the increasing number of remote sensing satellites. Reductions in computer costs and better miniaturization will also lead to more airborne sensing from smaller aircraft. Scientists need efficient handling and management of data to be productive. Due to the data volumes, it is foreseeable that no single institution will be willing to store all available data at its location. Only those data that are used and collected by an institution will be stored locally. To share data efficiently, organizations and institutions must abide to a mutually agreeable data standard. However, this is a difficult problem when data sharing is extended to international communities because of commercial value of the data, the resources, and fear of loss of sovereignty. SEIDAM supports the data formats supported by PCI Geomatics EASI/PACE image processing software for images and by the Environmental Research Systems Institute's (ESRI) ARC/Info software for GIS files. Data access must be efficient and reliable and remote access via high speed communication networks is an essential task. SEIDAM's ATM and high speed Ethernet communication networks are able to handle the transfer of large volumes of data for image and GIS browsing, queries and data transfers. For example, in Victoria, all government offices are linked through the Metropolitan Area Network, a fibre optic network for high-speed data exchange. The SEIDAM ATM Network links the BC Information Systems Technology Agency, BC Ministry of Environment, Lands and Parks, RNet, and the Universities of Victoria, British Columbia, and Ottawa.

Metadata

SEIDAM's Metadata Management System (SMMS) handles the following metadata issues:

1) provides trackable historical and legacy data records;
2) standardizes input, ingest, storage and output data formats;
3) implements a standard set of metadata manipulation tools;
4) provides separate image and GIS metadata facilities;
5) protects data from unauthorized distribution.

Metadata are vital to the successful and reliable use of data stored in the information system. SEIDAM contains an image metadata database and a GIS metadata database. Both metadata databases conform to the US standard set by FGDC. The metadata databases are implemented as textual flat files and are automatically updated as data are ingested and used in SEIDAM.

Scientific Visualization and Generalization

Visualization is an important element of a system such as SEIDAM and assists in comprehending complex results from image and GIS data fusion and analysis. Innovation is necessary in human-data interaction consistent with an era pervaded by the world-wide-web and the internet and the coming commercialization of a 'lifelike simulated reality'. Development in visualization functionality must parallel the expanding visual literacy of the population. The GIS challenge is to link the database entities with time attributes to an animation or video presentation, in order to satisfy the user's information needs, and provide cognitive stimulation through which new information may be obtained.

The support of true real time animation requires several technologies to be used or developed including:

1) high bandwidth transfer for large data volumes;
2) extraction of relevant information in real time from multi-temporal and multi-sensor imagery;
3) elimination of radiometric, geometric, and other systematic differences in multi-sensor and multi-temporal data;
4) co-registration of large data sets to facilitate dynamic visualization;
5) implementation of mass data storage capability for real-time access and data interchange;

6)   creation of high speed algorithms to effect real time perspective changes and presentation;

7)   a dynamic windowing system to select areas covered by multiple or part of the images in the data archive;

8)   an interface to specifically address the visual response as subjective (user dependent) criteria.

Visualization is closely akin to GIS and map generalization, since in both cases we want to present the optimum amount of information to the user or decision-maker. As an example of user need for map generalization, consider the Forest Inventory Program of the British Columbia Ministry of Forests. The mission of the Forest Inventory Program is to provide current, comprehensive, accurate, consistent and ecologically based inventories of the forest resources of British Columbia and to provide efficient access to the forest resources information. The Forest Resources Inventory Program now maintains approximately 7 000 georeferenced map sheets at a scale of 1:20 000. Approximately 6 700 are within crown-owned provincial forest boundaries. The objective of the program is to update maps on a two-year cycle to detect changes such as openings in the forest. A complete inventory is repeated on a ten year cycle.

There is also a need to be able to produce thematic maps at 1:250 000 scale or smaller that portray forestry information. The Ministry of Forests Inventory Branch maintains a single database with the ability to produce multiple scale representations from that database. Algorithms that generalize the 1:20 000 scale maps to a scale of 1:250 000 are available, and incorporate user-designed decision rules determining the generalized features. For example, one decision rule may be to define and generate "Old Growth". The user can review attributes of the database and design a specific definition for "Old Growth". SEIDAM makes use of the visualization tools found within PCI Geomatics' EASI/PACE and ESRI's ARC/Info, and also uses Advanced Visual Systems AVS and Research Systems Incorporated's ENVI software for additional image visualization.

Machine Learning and User Interaction

Analysis of remote sensing data has traditionally been performed using 3rd generation languages to code specific algorithms. These algorithms are usually input-output and computationally intensive due to their complexity and the amount of data associated with remotely sensed images. The operations of the algorithms and the programs supporting them yield results specific to the user's domain and interests. These include the application domain to which the data is applied, the remote sensing platform and sensor used, the desired input or output format, and other factors specific to the operational environment in which the user wishes to apply the results. These factors all augment the inherent computational complexity of the problem. This, combined with the number of algorithms required for a comprehensive image analysis system, can make an image analysis session in the native environment seem formidable to most users. For example, specification of parameters for the operation of the image analysis programs is through sequences of prompts presented to the user. These may seem repetitive, cryptic, or mundane. Certain prompts may be presented to the user with scientific or technical terms. Other prompts may assume a mathematical understanding of the algorithm. Users of one domain may understand some prompts which may seem foreign to users in other domains.

For these reasons we developed an automated program execution facility as a function of the expert shell. Expert systems at the initial stage of the decision path operate with a narrow scope of knowledge and perform the basic image analysis operations required to support decisions at a higher level expert system. These experts have knowledge of the prompting sequences of the image analysis programs. They interpret the information from the expert system within the context of the users application, which has been supervised by higher level experts. The initial expert system parses the prompts from the programs and provides responses based on the information already derived by the expert system. Image analysis programs are thus operated in a completely automated fashion using this facility. As a result, repetition of prompts is eliminated, cryptic prompts from the programs are handled consistently, and knowledge of the algorithms can be coded in the system experts in order to handle prompts which require a deeper understanding of the functions. The only prompting given to the user is through the expert system itself, and it is consistent for the user. Application domain experts ensure that requests for information outside the user's realm of knowledge are not seen. PAROT, the tool for creating expert systems in SEIDAM, can be learned by students with one week of training.

User Interface

The end user of any piece of computer software often judges its success or failure by the user interface alone. The expert shell (RESHELL) has a windows interface (Figure 2 middle) implemented at the Prolog level. The importance of providing the windows tools in Prolog is that many developers of expert systems maintain a standard user interface. Specific functions are performed automatically at the Prolog level, such as window sizing, widget selection, error messages and string to Prolog translation. A comprehensive rule and object editor has been implemented as part of this windows interface. Adapting the user interface to the expertise level of the user is included, as are cut-copy-paste operations, dynamically building multi-purpose prompt windows within a parallel-processing, non-deterministic prompting environment, window logging, and a design for the easy addition of future developments in the expert system.

EXAMPLE

As an example of the use of SEIDAM, consider the question of updating a forest inventory GIS file using remotely sensed data at the primary SEIDAM test site in the Greater Victoria Watershed District (GVWD) on Vancouver Island (Figure 2 bottom). The geographic boundaries of the study area are used as a mask to constrain the remotely sensed data to the geographic area of interest.

Initially, the geographically constrained image (Plate 16 top) is segmented. Classification on this segmented image helps to identify certain structural units such as forest at various stages of growth, rock outcrops, and areas of forest cover change. The result of the segmentation process is shown in Plate 16 (middle) in which structural units are clearly visible in the image as polygons. Next, classification produces the result shown in Plate 16 (bottom). In the final step of updating the polygon attribute database, the forest cover change polygons of Plate 16 are exported to the GIS data.

CONCLUSIONS

SEIDAM simplifies user access, data management, and data integration by embodying a distributed group of expert systems operating on several different computers networked together. SEIDAM includes capabilities for using data from several satellites, aircraft-borne instruments, and field measurements obtained on a sampling basis. SEIDAM is also useful as a tool to train operators in data fusion. In the context of forestry, SEIDAM has several benefits including: (1) increasing the use of remote sensing data; (2) reducing the reliance on staff with rare skills; (3) ensuring consistent accuracies for provincial and national resource databases; and (4) serving as a useful tool to train existing operators in the analysis of remotely sensed imagery. This training enables the operator to recognize forest objects of interest, such as harvested areas, areas of poor growth or poor regeneration, and areas under stress from environmental causes or insects. In a more general context, SEIDAM advances a number of computer science concepts using intelligent agents to access corporate databases, execute processes on distributed systems, and update remote databases.

SEIDAM uses a set of intelligent software agents to carry out a variety of tasks. For example, SEIDAM agents can automatically integrate a topographic GIS file with a digital forest cover GIS file, and remote sensing imagery such as from LANDSAT Thematic Mapper. Users can employ automatic classification algorithms and segmentation rather than carry out manual digitizing to digitize clear cut, and another set of software agents update a forest cover map given the newly digitized clear cut polygons. Development of SEIDAM requires expertise in several different fields: forestry engineering, database systems, GIS, remote sensing and digital image analysis. SEIDAM uses artificial intelligence technologies to perform both the management and processing of GIS and remote sensing data with a case-based reasoning system to assemble plans that can be executed to create integrated products (Goodenough *et al.*, 1998; Charlebois *et al.*, 1996a, b).

ACKNOWLEDGMENTS

We gratefully acknowledge support from Natural Resources Canada. Dr. Goodenough also acknowledges support from the Natural Sciences and Engineering Research Council of Canada. We thank our co-investigators in SEIDAM and NASA for their contributions to the project. Additional information and SEIDAM software can be found at http://www.aft.pfc.forestry.ca.

REFERENCES

1. Canadian Council of Forest Ministers (1995) *Defining Sustainable Forest Management: a Canadian Approach to Criteria and Indicators*, CFS Publication, Fo-75-3/4-1995E.
2. D. Charlebois, D.G. Goodenough, P. Bhogal and S. Matwin (1996a) Case-based reasoning and software agents for intelligent forest information management, in *Proceedings of the 1996 International Geoscience and Remote Sensing Symposium*, pp. 2302–2306. Lincoln, Nebraska, USA.

3. D. Charlebois, D.G. Goodenough, S. Matwin, A.S.P. Bhogal, and H. Barclay (1996b) Planning and Learning in a Natural Resource Information System, in *Proceedings of Canadian AI*, pp. 187–199. Toronto ON, Canada.

4. Federal Geographic Data Committee (1997) *Framework Introduction and Guide*. (Washington DC: FGDC).

5. M. Goldberg, D.G. Goodenough, M. Alvo and G. Karam (1985) *IEEE Trans. Geosci. Remote Sens.*, **73**, 1054–1063.

6. D.G. Goodenough, A.S. Bhogal, A. Dyk, R. Fournier, R.J. Hall, J. Iisaka, D. Leckie, J.E. Luther, S. Magnussen, O. Niemann, and W.M. Strome (1999) Earth observation for sustainable development of forests – A national project, in *Proceedings of the 1999 International Geoscience and Remote Sensing Symposium*, pp. 2101–2104. Hamburg, Germany.

7. D.G. Goodenough, D. Charlebois, A.S. Bhogal and N. Daley (1998) An Improved Planner for Intelligent Monitoring of Sustainable Development of Forests, in *Proceedings of the 1998 International Geoscience and Remote Sensing Symposium*, July 6–10. Seattle, Washington, USA.

8. D.G. Goodenough, M. Goldberg, G.W. Plunkett, and J. Zelek (1987) *IEEE Trans. Geosci. Remote Sens.*, **GE-25**, 349–359.

9. S. Minton, J. Bresina. and M. Drummond (1994) *J. Artificial Intel. Res.*, **2**, 227–262.

10. UCGIS (1996) *Cart. Geog. Inform. Syst.*, **23**, 115–127.

11. World Commission on Environment and Development (1987) *Our Common Future*, 400 pp. (Oxford: OUP).

Section 3

Analysis and Management of Land Use

11. MAPPING BIODIVERSITY FOR CONSERVATION AND LAND USE DECISIONS

Michael J. Conroy

INTRODUCTION

Decision making in conservation biology is frequently a complex process, involving the balancing of conflicting objectives, and the processing of vast quantities of information. Geographic information systems (GIS), statistics, models, and other techniques can be extremely useful tools in the decision process. Examples of decisions to which these techniques might be applied include reserve design, forest cutting, and species reintroduction. In each, technology can be a helpful adjunct to the decision process, but does not provide that process. GIS and spatial models may be useful in displaying alternate configurations of a reserve design, but only if the decision maker has a clear objective (e.g., maximizing species diversity; ensuring persistence of particularly valued species or habitats). Likewise, both the data and the models are subject to statistical and other errors, or may be dependent on untested assumptions. The extent to which this information may be useful to the decision maker will depend on many factors, all of which depend on how they will be used in a decision process.

STEPS IN THE DECISION MAKING PROCESS

Goals and Objectives

The clear delineation of goals and objectives is essential to sound decision making in conservation, and provides direction for management and a means for evaluating the success of management actions. *Goals* are broad statements of a desired consequence resulting from accumulated conservation decisions, *e.g.*, "maintain a desired balance between biological diversity and production of timber." *Objectives* are mathematical expressions of the goal, and should be predictable from existing information and models and be measurable from field or other data. Objectives also must quantify tradeoffs (for example, among multiple resources) and constraints that are imposed on the decision-maker, and identify a time frame over which the objective is to be achieved. One way to achieve a balance among conflicting objectives is to allocate different resource components by space or time, so as to achieve each independently and without mutual conflict. However, this approach must take into account spatial dependencies that may exist among spatial components of the system. For instance, a population of nesting songbirds may require large blocks of contiguous forest or riparian habitat, in order to provide sufficient opportunities for dispersal of young. A timber cut interposed between two otherwise suitable (from a local population standpoint) habitats, may render both unsuitable.

Decision Alternatives

The key to the decision process, is the proper identification of *decision alternatives, i.e.*, the range of possible management actions that can be taken by the decision-maker. Obviously, these will be constrained by many factors, such as land ownership, the extent of the decision maker's proprietary or legal control, applicable laws and regulation, cost, feasibility, and time. In practice, the decision-maker may find that the actual range of decisions is a small subset of those that conceivably could be taken. Thus, the manager of a forest reserve might desire to exclude roads from the reserve in order to enhance the biodiversity objective, but is unable to do so because of legal easement and other arrangement prescribing such access.

In addition to such constraints and considerations, the decision-maker must consider spatial and temporal aspects to decision making, and the extent to which particular decisions, once made can actually be implemented in the field. For instance, prescribed fire is frequently used to restore and maintain certain plant and animal communities, particularly where it is impractical to rely upon wildfire. The manager must not only decide how much of an area to burn, but how frequently particular areas should be burned, and what the spatial allocation of fire should be at any given burn season. All of these aspects will depend on the resource objectives, and on how decisions will (or may) affect the system so as to lead toward or away from the objectives.

Knowledge Base and Models

In most cases there will be at least some scientific basis for asserting that some decisions should lead toward achieving an objective, whereas others should lead away from that objective. This may include data from the specific area under management or consideration (*e.g.*, soils, vegetation, and faunal and floral surveys), information and theory from similar habitats, and expert opinion. This knowledge may be used to construct formal, mathematical models of system behavior, but even *in lieu* of such models managers frequently will have a mental "model" of expected system response to particular decisions. A mathematical model has the obvious advantage of enabling quantitative, testable predictions. In addition, a mathematical model can be encoded and combined with other databases and models, for example in GIS, allowing rapid display of alternative decision scenarios.

Models may be capable of making spatially explicit predictions, for example point locations of animals through time. Alternatively, predictions may be of population summary statistics such as abundance, perhaps stratified by habitats or other units of space; of total abundance only; or of presence or absence of animals, without regard to numbers. The appropriate degree of spatial and temporal resolution of predictions depends on the specific application of the model, and whether the discrimination among decision scenarios to find an optimum objective result requires this resolution. In addition, model resolution will depend on the capacity of the modeller to secure the necessary data to estimate model parameters and validate the model (Conroy *et al.*, 1995). Often simpler models are preferable, because there is no hope of obtaining data appropriate for more complex modeling.

This last point raises the issue of the role of surveys and monitoring programs in decision making. Monitoring has at least four roles, all directly linked to enhancing the ability to make optimal resource decisions. First, monitoring can establish a benchmark for current conditions, from which projections can be made (perhaps using models) as to the likely impacts of alternate management decisions. Second, monitoring can be used to assess whether the stated objective has been reached, or whether the system appears to be departing from the objective. Third, monitoring can be used to predict the impact of alternative management decisions on future system states and objective values. Finally, monitoring can provide adaptive feedback for improving system knowledge and future decision. In adaptation (and adaptive management) predictions are made about the consequences of various possible management actions, but now one explicitly considers the possibility that these may be at least partially wrong, because of an inadequate understanding of how the system works. We nonetheless need to make a decision, so we will do so given our best current understanding (predictions) about how the system will respond, meanwhile collecting monitoring data to examine our predictions, leading to improvement in our future predictive capacity, and to better future decisions.

Optimal Decision Making Under Uncertainty

Occasionally the results of a conservation decision are highly predictable. For example, a reserve boundary drawn about a habitat of a certain current composition will have a more-or-less predictable appearance in 10 yr, given assumptions about rates of succession and the likelihood of intervening events such as hurricanes or wildfire. However, in most cases predictions about the impact of management will be inexact. This *uncertainty* in predictive ability must be considered in decision-making, and is from several sources. First, natural systems are subject to intrinsic uncertainty, arising from environmental variability, demographic processes, and other factors. Second, usually the system has not been measured exactly, but is subject to sampling and measurement errors. Third, typically the implementation of decision is not under complete control (*e.g.*, a prescribed burn may affect a larger or smaller area than intended). Finally, typically systems are incompletely understood, so that potential responses to management are not fully predictable, even taking into account the above sources of uncertainty.

Each of these sources of uncertainty contributes to the "noise" in the decision process, and will weaken the connection between a decision and the desired objective result. Statistical uncertainty can be reduced (but not eliminated) by increasing the intensity of sampling and monitoring efforts. Environmental or demographic uncertainty cannot be controlled, but must be taken into consideration in decision making. The last source of uncertainty, *structural uncertainty* due to a fundamental lack of understanding of system processes, requires special attention. Like the other sources of uncertainty, structural uncertainty cannot be ignored, but must be taken into account in decision making (*e.g.*, Lindley, 1985). One way is to consider multiple, alternative models for describing the system's potential response to decision alternatives. For instance, suppose that two possible reserve designs (A and B) are under consideration. One (Model 1) of community dynamics predicts that under A biodiversity objective will be 50% higher than under B. However, we are unsure as to the validity of this model, and so entertain a model (Model 2), having different parameter values, structural assumptions, or both, that predicts no difference with respect to the biodiversity objective. Assume that these are the only two models available to us, and that neither can be disconfirmed by comparison to data. At this point, it would be tempting to pick the model (say, Model 1) that most agrees with our prior notion of how the decision *ought to* affect the system. However, unless we are *certain* (and we are not) that this is the correct model, this is a risky way of making a decision. At the very least we should admit the possibility that *either* model may be reasonable, and make a decision accordingly. We may wish to give one model greater weight in this process, if we have objective reasons for doing so. In many instances, we will have no reason to give one model more weight; indeed, it is common for data to be incapable of discriminating between the models (in a hypothesis testing framework). Fortunately, structural uncertainty is reduced, and decision-making improved, by *adaptive resource management* (ARM; Walters 1986). ARM involves making management decisions that appear to be optimal, based on current information (model weights). Each model will make a different prediction about the impacts of the decision, and these predictions can be compared to post-decision observations collected through monitoring and research programs. The model whose prediction best matches these observations will receive more weight in the next round of decision making. This procedure repeats through time (or, as discussed below, space) with the cumulative result that both knowledge and decision making improve through time.

The Role of Science in Decision Making

Decision making in natural resource management can be extremely contentious, because of the high value placed on outcomes by various stakeholders. I wish to make clear the distinction between disagreements over *outcome* (*i.e.*, the objective), and uncertainty about scientific processes (often represented by alternative models). The former is fundamentally the realm of the social sciences (policy, economics, *etc.*), whereas the latter deals with our ability to understand, and thus to varying degrees effect control of ecological systems. Unless this distinction is made, uncertainty (in the latter sense) may be exploited by stakeholders having very different (and often, uncompromising) views regarding goals and objectives. When decisions are contentious, these frequently become entangled, and "science" becomes a smokescreen for political agendas. In the controversy over appropriate forest management for the endangered spotted owls (*Strix occidentalis*), scientific uncertainty exists over the extent of old-growth forest required to sustain owl populations. Not coincidentally, stakeholders holding different primary interests, caricaturized as "jobs" *vs.* "old-growth", tended to favor the scientific viewpoint most favorable to their interests. As seen above, uncertainties about ecological processes can be dealt with objectively, and under ARM can be reduced by the integration of optimal decision making, monitoring, and adaptive updating. However for ARM (or any other objective decision process) to succeed, disagreements over goals and objectives must be first be resolved in the domain of social, legal, and political processes.

CASE STUDY: FOREST MANAGEMENT AT PIEDMONT NATIONAL WILDLIFE REFUGE

I illustrate these ideas with an example of a natural resource decision problem at a wildlife refuge administered by the U.S. Fish and Wildlife Service (FWS). The Piedmont National Wildlife Refuge (PNWR) is located in central Georgia and comprises approximately

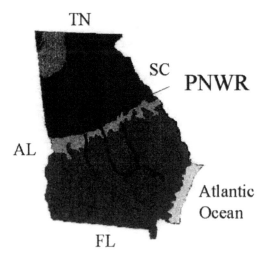

Figure 1. Location of the Piedmont National Wildlife Refuge in the southern Piedmont of Georgia.

14,000 ha of mostly second growth forested habitats (Figure 1). The refuge was established in 1939 following the reversion of much agricultural land in the region to forest as a result of economic and environmental events of the 1930–40's.

Goals and Objectives

Management of PNWR is directed toward a number of objectives including recreation and wildlife conservation. The refuge is legally mandated to manage for recovery of the endangered red-cockaded woodpecker (*Picoides borealis*; RCW), a cooperative breeder. Populations of RCW are encouraged by artificial nest inserts and by silviculture. Specifically, silviculture favors older (75+ yr) stands of loblolly pine and shortleaf (*Pinus taeda*, *P. echinata*). Suitable nesting and foraging habitats are maintained by thinning (removal of merchantable pine trees) and prescribed burning in order to maintain the open, savanna-like conditions thought to be favourable to RCW (Richarsdon *et al.*, 1998). However, these conditions may be at least locally unfavourable to other species that depend on denser cover, particularly of understory and midstory shrubs and trees, such as the shrub-nesting wood thrush (*Hylocichla mustelina*; WOTH), whose number have been declining for decades. Other species of wildlife are important to managers and stakeholders at PNWR, but I will concentrate on the objective tradeoff between RCW and WOTH. For illustration I took, as simple indicators of population "health", the finite rate of population increase (λ) for each species.

$$\text{Objective} = \lambda_{\text{WOTH}} - (\lambda_{\text{RCW}} - 1) \tag{1}$$

This objective function has the desirable properties that: 1) it is maximized when both λ_{RCW} and λ_{WOTH} are at maximal values (both populations increasing); and 2) it is bounded above by zero when RCW populations are non-increasing ($\lambda_{\text{RCW}} \leq 1$).

Decision Alternatives

Here I focus on decisions most related to the objective function (1): the timing and aerial extent of timber harvests (*i.e.*, regeneration cutting), thinning, and prescribed burning. The scheduling of these activities is made as part of a refuge planning process, typically over a 5–10 yr time horizon; refuge plans are revisited and revised periodically (*e.g.*, at about 5-year intervals). These decisions are implemented on an annual basis, as forest stands reach the age or stocking deemed appropriate for intervention. Stands are organized spatially into administrative units of approximately 400 ha each (compartments). Compartments are further grouped into about five blocks of seven compartments each (35 total refuge compartments), and silvicultural activities rotated among these blocks so as to complete management prescriptions over an approximately 8-yr period.

Knowledge Base

The knowledge base used to conduct management at PNWR is based on published and unpublished studies at PNWR and similar habitats in the Southeast, and the accumulated experience of refuge staff, as well as guidelines established to meet regional objectives such as the recovery of RCW (Richardson *et al.*, 1998). This knowledge is reflected in the structural forms of the models developed for predicting changes in habitats and animal populations in response to management. In some cases specific values for model parameters were available from ongoing or previous studies at PNWR or elsewhere in the Southeast (*e.g.*, from similar areas). In others, we had little choice but to make educated guesses as to parameter values. In either case, alternative model structures, parameter values, or both are incorporated in the decision process.

Forest growth models

Currently our forest growth model (Powell, 1988) is limited to the pine stands under active management, with other (*e.g.*, hardwood) stands treated as static. We defined three forest age classes: 0–24 yr (*P1*), 25–65 yr (*P2*), and >65 yr. This last class was further divided into two sub-classes: *P3*, overstory conditions (principally, density of stocking) not suitable as RCW nesting habitat; and $P3_{RCW}$, managed (through thinning) to maintain conditions suitable for RCW nesting. We used a projection model similar to Reed and Errico (1986) to model overstory dynamics as the transition of acreage of pine habitat between age classes per year. In this chapter I focus on the shorter-term dynamics of understory cover important to RCW and WOTH. Dense shrubby vegetation and small trees are thought to be important to WOTH for nesting, but disadvantageous to RCW (Lang 1998, Powell 1998). I modeled dynamics of *P3* habitats which previous data had shown to be suitable habitats for WOTH (given the presence of dense understory vegetation) and which are being managed for RCW nesting and foraging habitats by the removal of this vegetation by cutting and prescribed burning. The simple model for understory vegetation density is based on measurements of understory vegetation following prescribed burning, and describes vegetation density in four density classes: *U1*, 0–10%; *U2*, 11–33%; *U3*, 24–67%; and *U4*, 66–100% understory cover, respectively. Previous analyses lead us to predict that in the absence of further prescribed burning, vegetation would re-grow so as to achieve the *U4* density class with 8 yr. Prescribed burning affects dynamics by re-setting vegetation growth to zero, for that portion of the area affected by the burn.

Animal population models

The animal population models depend fundamentally on information provided by the vegetation dynamics models, in determining the amount of suitable habitat available for each species. However, habitat "suitability" is not in itself of interest, but rather the abundance and growth rate of RCW and WOTH populations, under the assumption that these quantities are observable at some scale of resolution. Our population models thus sought to predict changes in numerical abundance of each species, as a function of: (1) the current population state (numerical abundance, plus possibly demographic and spatial structure); and (2) habitat conditions, with these being at least partially under management control. Thus, a generic form of our population models is:

$$N_{t+1} = N_t \, \lambda_t \, (H_t, \, z_t) \tag{2}$$

where N_t is abundance at time t (again, possibly stratified by demographic classes or space), λ_t is the finite rate of increase from time t to $t + 1$, H_t is a vector of habitat conditions at time t, and z_t is a vector of random effects at time t. The essential aspects for each species are summarized as follows:

Woodpeckers. We modified a model by Heppell *et al.* (1994) to incorporate the effects of vegetation succession and management on RCW dynamics. First, suitable habitats for nesting (H_n) and foraging (H_f) are defined as $H_f = P3(t)*U1(t)$ and $H_n = P3'(t)*U1'(t)$ where $U1(t), U1'(t), P3(t)$, and $P3'(t)$ are the understory vegetation density and pine acreage described in the previous section. Our model predicts increasing juvenile and adult survival as the amount of foraging habitats in $U1$ and $U1'$ increases, according to

$$y_i = 0.8 + 0.4*H_i/min_i, \qquad \text{if } H_i < min_i;$$
$$y_i = 1.2, \qquad \text{if } H_i \, min_i, \, i = f, \, n, \tag{3}$$

where y_f is a multiplier affecting survival rates, and y_n is a survival affecting the reproduction rate. Minimal habitat thresholds of $min_f = 100$ ha and $min_n = 20$ ha, respectively, are based on guidelines in Richardson *et al.* (1998). This model was then used to predict the numbers of breeding and non breeding RCW at the next census time ($t + 1$), as a function of those present at time t, habitat conditions at t, and management (removal or non removal of vegetation). It describes the population without respect to spatial stratification or movements, but can be extended to the more general case in which population states are observed on a stratified (*e.g.*, by forest stand or compartment) basis.

Wood thrush. In contrast to RCW, WOTH have fairly general habitat requirements. Previous studies (*e.g.*, Powell, 1998) have indicated a preference by WOTH of forested habitats containing a mixture of large canopy trees and a well developed midstory suitable as nesting substrate. In addition, WOTH generally prefer to nest near streams or moist areas, in part because they incorporate mud into the building of nests. Thus WOTH habitat requirements may conflict with those of RCW, which as noted above depend on the relative absence of understory-midstory structure. Evidence from fieldwork at PNWR (Powell, 1998) suggests that WOTH may be fairly tolerant to habitat modification directed toward RCW, in part because of their proclivity to move over large areas during the breeding season, and between nesting attempts, in search of suitable nesting and foraging habitats. We described the impact of habitat modification on WOTH populations by

$$\lambda(H_s/H) = 1.5\text{--}2\ H_s/H \tag{4}$$

where H_s/H is the proportion of all habitat at PNWR considered as suitable. This relationship was then used to predict the number of WOTH at census time $t + 1$, as a function of WOTH numbers and habitat conditions at t, and intervening management (removal or non removal of vegetation). As with RCW, spatial aspects will be incorporated, principally to reflect minimal area constraints of suitable habitats; work by Powell (1998) suggests that WOTH are capable of moving over large distances during the breeding period, and thus other spatial aspects (*e.g.*, juxtaposition of habitats) are not considered important.

Survey and Monitoring Data

Monitoring data at PNWR are required to: (1) evaluate the current state of the forest vegetation and animal populations of interest so that appropriate management decisions can be made; (2) determine the extent to which the objective function (equation 1) is being met; and (3) to provide a basis for evaluating model predictions and adaptively incorporating new knowledge about system processes. In the simplified forest growth model, only the distribution of forest area across age classes is required, and is available from PNWR records for all refuge compartments. Density of midstory and overstory vegetation are obtained from annual surveys from transect lines placed in a sample of forest compartments (Powell 1998). Additional measures are taken of forest canopy, tree basal area, site quality, and species composition, which may be used in subsequent, more sophisticated modeling of forest overstory and understory growth.

Visual counts were made of RCW and WOTH to determine total numbers at each site and for RCW the number of breeders and non-breeders. These data may either be aggregated to a total refuge basis (as in the simplified model presented here), or considered on a compartment or stand basis (for spatially structured modeling; see below). Estimates of survival, fecundity, and stage transition for RCW are based on Heppell *et al.* (1994) and Stevens (1995). Estimates of the relationship between WOTH population growth and management for RCW are based on Powell (1998). This study showed no effect of management treatments on WOTH parameters. Nonetheless, the nature of the treatments meant that the probability of statistically significant differences was low, and we have chosen to consider an alternative model in which total removal of understory and midstory (as would occur under much more aggressive management than applied so far) would negatively influence WOTH growth rates.

Effects of Structural Uncertainty on Decision Making

Even for the above simple model of the system and management, data are inadequate for discrimination between alternative, plausible responses to management. This uncertainty is likely to be more profound for the response of bird populations to management, whereas vegetation response is relatively (but not perfectly) predictable. Here I illustrate the potential implications of model uncertainty on decision making by conditioning on a situation in which the responses of vegetation and RCW populations are perfectly represented by our models and that environmental, demographic, and statistical sources of uncertainty are negligible. I focus on structural uncertainty in the RCW response to management, by the creation of a set of two alternative models. In the first (Model 1), the parameter structure and values for both RCW and WOTH is as indicated above. Management is simplified to consist only of the removal of understory and midstory vegetation, with no removal resulting in all acreage in both $P3$ and $P3'$ occurring in the densest classes ($U4$, $U4'$), and total removal resulting in all acreage in both $P3$ and $P3'$ occurring in the densest classes ($U4$, $U4'$). For simplicity in this illustration I assume that management is able to maintain vegetation densities in the given states with total control, and that over the time horizon of interest initial areas of 200 and 20 ha for $P3$ and $P3'$, respectively, remain constant, thereby removing any consideration of vegetation dynamics. Figure 2 illustrates the response of this modeled system to various levels of vegetation removal (0–100 %) in terms of lambda for RCW and WOTH and the value of the objective function (1). Maximum values for RCW occur under 100% vegetation removal, and the reverse is true for WOTH (Figure 2, top and middle), resulting in an optimal objective value at intermediate levels of vegetation removal in $P3$ but 100% removal in $P3'$, due to the relatively small contribution of the latter type to WOTH habitat.

Under the alternative assumption (Model 2), vegetation removal in nesting habitats has the impact on RCW as specified in (3), but now removal in foraging habitats is assumed not to benefit RCW ($y_f = 1.0$). Under these assumptions a very different picture emerges (Figure 3). Here, the predicted response for WOTH remains the same, no relevant assumptions having changed, but now RCW are influenced only by removal in Figure 3 (top and middle). The impact of these different assumptions on the objective function is profound, with optimal management now occurring under total removal of vegetation in $P3'$, but no removal in $P3$, reflecting the negative impact on WOTH of unnecessary removal of vegetation in this type. The point here is not that this (or either) model is "correct" in any absolute sense (*e.g.*, Conroy 1993), but rather that uncertainty with respect to this specific component of system response to management can lead to very different, apparently optimal decisions. This emphasizes the importance of: (1) taking structural uncertainty account in decision making; and (2) the value of reducing this component of uncertainty, as will be discussed in the next section.

Structural uncertainty can be incorporated into decision making by computing the value of the objective function for each possible decision, for each model, and constructing a weighted average as

$$E(Obj_j) = (Obj_j|m_1)p(m_1) + (Obj_j|m_2)p(m_2) \tag{5}$$

where $E(Obj_j)$ is the expected value of the objective function, given decision j, and taking into account structural (model) uncertainty, $Obj_j|m_i$ is the value of the objective function for decision j assuming model i is correct, and $p(m_i)$ is the probability that model i is correct. Assuming that some measure of relative belief (ranging from 0 to 1) in each model is available, expression (1), which can be generalized to more than two alternative models and to continuous alternatives, dictates that the optimal decision, d_j is that which maximizes $E(Obj_j)$. Clearly, to the extent that alternative models lead to different decisions, as they

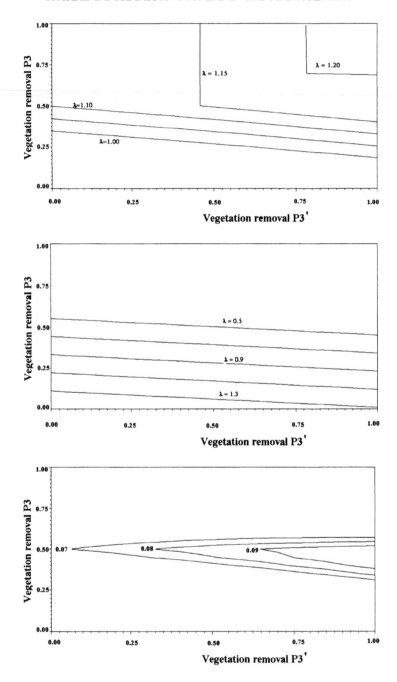

Figure 2. Predicted population growth rates for red-cockaded woodpeckers (top) and wood thrushes (middle), and the value of the objective function (bottom) in relation to levels of prescribed burning, under assumed model structure (*Model 1*) in which wood thrush respond negatively to increases in prescribed burning. Contour lines represent combinations of vegetation removal in P3 and P3' resulting in specified levels for population growth (λ) or value of the objective function (see text).

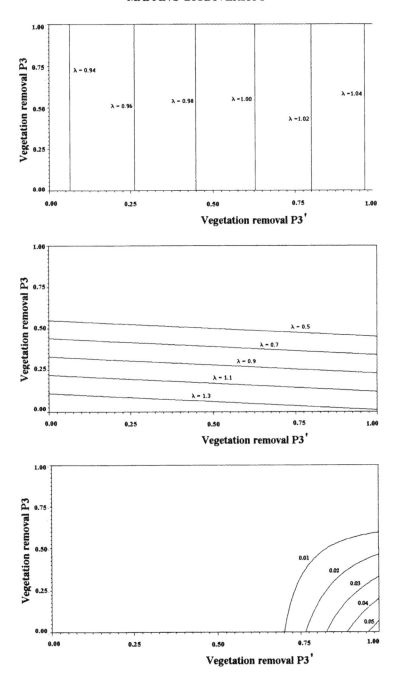

Figure 3. Predicted population growth rates for red-cockaded woodpeckers (top) and wood thrushes (middle), and the value of the objective function (bottom) in relation to levels of prescribed burning, under assumed model structure (*Model 1*) in which wood thrush do not respond to increases in prescribed burning. Contour lines represent combinations of vegetation removal in *P*3 and *P*3' resulting in specified levels for population growth (λ) or value of the objective function (see text).

do in the simplified examples in Figures 2 and 3, it will be important to reduce uncertainty among plausible alternative models.

Adaptation

Given a current set of habitat and population states, a decision (*e.g.*, the degree of vegetation removal), and alternative models, one can predict future states of the system following implementation of the decision (presumably chosen to be optimal, given current knowledge). Assuming that these states can be observed at future times, it is straightforward to compare the predictions under each model to the observed outcomes (data), and modify model weights by an application of Bayes' Theorem:

$$p(m_1|\text{data}) = p(m_1)L(m_1)/[p(m_1)\ L(m_1) + p(m_2)L(m_2)] \qquad (6)$$

where $p(m_1)$ is the prior probability of model *i*, before comparison of model predictions to data, $p(m_i|\text{data})$ is the posterior probability of model *i*, after comparison of predictions to data, and $L(m_i)$ is the statistical likelihood of the data under model *i*. Expression (6) can be generalized to allow for any number of alternative models, and provides a means for sequentially updating the probability weight for each model; ideally, weights will approach one for one of the models and zero for the other(s).

Spatial structure

The above representation of the system at PNWR incorporates the effects of habitat management though simple demographic models, and allow specific predictions of population growth, based on simple habitat and population state variables. A more realistic representation of the system will incorporate the fact that management actions can be arranged in various ways on the landscape of PNWR, resulting in different spatial arrangements of habitat conditions at any time *t*. The spatio-temporal nature of management and vegetation response can be captured by a fairly straightforward extension of the vegetation dynamics models to allow for spatial stratification. Under the assumption that stand and compartments are independent, reasonable with respect to the impacts of cutting and burning actions, vegetation dynamics can be modeled by the simple aggregation of separate stand- or compartment-level models. Explicit consideration of spatial structure is only needed to the extent that these influence the objective function (3), either directly, *e.g.*, by imposing logistic costs or constraints or indirectly, through the influence of spatial structure on demographics. The first concern is addressed to some extent by the practice of organizing management into groups of stands (compartments) receiving similar management action in any year, so that efforts are not randomly dispersed across the landscape. With respect to the second issue, spatial arrangement of habitat states could be important, to the extent that stand or compartment-level populations are demographically interdependent. At present, we believe that spatial representation is relatively unimportant for WOTH, in that virtually the entire PNWR may be available to WOTH during the course of a breeding season (Powell, 1998).

However, RCW dynamics may be critically dependent upon spatial structure, because RCW occur in low numbers, are highly social, and have specialized habitat requirements. The extent to which spatial dependencies are important to RCW is not known, and alternative possibilities seem worth considering, particularly as these relate to management. Plate 17 illustrates three possibilities in this regard; all assume that a potential breeder is dispersing from a natal colony to one of several suitable and currently unoccupied habitats. In the first scheme (Plate 17 top), dispersal is completely independent of either the distance to suitable habitats, or the quality of habitats along the movement path. In this scenario there clearly is little need to consider spatial structure in modeling. In the second scheme (Plate 17 middle), dispersal depends on distance but not on habitat quality, and in the third (Plate 17 bottom), dispersal depends both on distance and habitat quality. If the population is behaving as in Plate 17 (middle or bottom), then the spatial arrangement of habitat management may be critical. Two alternative management scenarios, one with a highly dispersed and fragmented habitat outcome, and the other with quality habitats close and/or spatially connected, would be expected to generate very different outcomes with respect to the RCW objective. Discrimination among these alternative models of RCW movement is important, either to avoid undesirable limitations on RCW growth, or to unnecessarily constrain other objectives (WOTH, timber production, etc.). Again, model uncertainty can be handled by: (1) evaluating the objective outcome under each model and constructing a weighted average for each decision; and (2) evaluating predictions from each model in an adaptive management framework.

SUMMARY

Data, GIS, and models may be useful tools to aid in decision making, but in the absence of a sound decision process they will contribute little. First, decisions must be directed toward established objectives; these often will require an explicit tradeoff (or spatial-temporal allocation) between conflicting natural resource goals. Second, decisions should be based on the best, current understanding of the likely system responses to alternative decisions. In most cases this must include explicit recognition that scientific uncertainty exists, and that the outcomes following the decisions may or may not lead toward a desired objective. Third, the systems under management must be monitored, and predictions made (through models where possible and appropriate), at a level of resolution consistent with: (1) capturing of the essential, possible responses to management; and (2) feasible and practicable monitoring programs (Conroy et al., 1995). Finally, decision making ordinarily can and should proceed in an adaptive manner, wherein current decisions are made based on the best current knowledge, but monitoring is conducted so as to provide feedback into reducing system uncertainty, thereby improving future decisions. Following these or similar procedures for conservation decision making will be scientifically defensible, provide a better allocation of scarce monitoring and other resources, and provide a better linkage between science and management.

REFERENCES

1. M.J. Conroy (1993) *Trans. N. Am. Wild. Nat. Resour. Conf.*, **58**, 509–517.
2. M.J. Conroy, Y. Cohen, F.C. James, Y.G. Matsinos, and B.A. Maurer (1995) *Ecol. Appl.*, **5**, 17–19.
3. S.S. Heppell, J.R. Walters, and L.B. Crowder (1994) *J. Wildlife Manage.*, **8**,479–487.

4.   J.D. Lang (1998) Effects of thinning and burning in pine habitat on wood thrush nesting, fledgling dispersal, and habitat use, MSc. Thesis, University of Georgia, Athens.

5.   D.V. Lindley (1985) *Making Decisions*, Second edition. (New York: Wiley).

6.   L.A. Powell (1998) Experimental analysis and simulation modeling of forest management impacts on wood thrushes, *Hylocichla mustelina*, Ph.D. Dissertation, University of Georgia, Athens, 198 pp.

7.   W.J. Reed, and D. Errico (1986) *Can. J. Forest Res.*, **16**, 266–278.

8.   D. Richardson, R. Costa, and R. Boykin (1998) Strategy and guidelines for the recovery and management of the red-cockaded woodpecker and its habitats on national wildlife refuges. (Atlanta, GA: U.S. Fish and Wildlife Service Region 4).

9.   E.E. Stevens (1995) Population viability considerations for red-cockaded woodpecker recovery, in *Red-cockaded woodpecker: recovery, ecology, and management*, edited by D.L. Kulhavey, R.G. Hooper and R. Costa, pp. 227–238. (Nacogdoches, Texas: Stephen F. Austin University).

10.   C.J. Walters (1986) *Adaptive Management of Natural Resources*. (New York: MacMillan).

12. CONTRIBUTION OF SPATIAL INFORMATION AND MODELS TO MANAGEMENT OF RARE AND DECLINING SPECIES

Virginia H. Dale, Anthony W. King, Linda K. Mann
and Tom L. Ashwood

INTRODUCTION

As habitats for rare and declining species are reduced on private and federal lands, there is mounting pressure to manage lands in such a way as to protect these natural resources. Ecological models can be used to explore ways that activities on the land can be conducted with minimal impact on natural resources. These models can also be used to evaluate the types and sensitivity of ecosystems and organisms at risk. They can help determine whether, under certain temporal or spatial restraints, land uses can be conducted in such a way that resources are not put at risk. The advantage of using ecological models is that both land-use goals and ecological risks can be simultaneously explored without jeopardizing either of these objectives.

The examples of spatially-explicit ecological models we present deal with land-use decisions typically made by the U.S. Department of Defense (DoD) on their installations. DoD must meet its mission needs while at the same time conserving ecological resources; however, DoD missions frequently require training or testing on land areas that may jeopardize these resources. In the United States, the Endangered Species Act requires that federal land actions protect threatened and endangered species, and the National Environmental Protection Act, the Clean Water Act, and the Clean Air Act require protection of environmental resources. Thus, military training and testing must conserve natural resources protected by such federal regulations.

The conservation task is not a minor one. Since the DoD lands have been largely protected from development and other disturbances, and because large areas were set aside to buffer military missions, the DoD lands now contain as many as 80% of the threatened and endangered species that are found on all federal lands in the U.S. (Leslie *et al.*, 1996).

The purpose of this chapter is to illustrate how we developed and applied ecological models that can provide land managers with information about the impacts that land use and management activities have on natural resources and, especially, rare species. The general approach uses a suite of ecological models that can predict where natural resources occur under various land management strategies and can identify features that are at risk under particular land-use scenarios. The types of ecological models that are useful to land managers include habitat models, population models, transition models, and regional models. The appropriate model should be selected depending on the issue at hand. The range of models discussed in this chapter and the context for their application is shown in Table 1. The models use readily available *in situ* and remotely sensed data and are

Table 1. Organization of ecological models for management of rare and declining species described in this chapter.

Model Types	Model Description	Organisms and Habitats	Sites	Land Management Issues
Habitat models	Soil-based GIS overlays	Threatened calcareous habitat	Fort Knox Military Preserve (Kentucky)	Identification of critical habitats
		Lupine habitat as a host for Karner blue butterfly	Fort McCoy (Wisconsin)	
Population models	Patch characteristics linked to demographic and survivorship functions	Henslow's sparrow nesting habitat	Fort Knox (Kentucky) Fort Riley (Kansas)	Managing threatened species on a single installation
	Habitat suitability linked to demographic functions	Karner blue butterfly habitat	Fort McCoy (Wisconsin)	
Transition matrix model	Probability of change from one landscape type to another as a result of management	Lupine habitat as affected by wheeled and tracked vehicles	Fort McCoy (Wisconsin)	Predicting results of land management actions
Regional model	Cluster characteristics and dispersal functions in a landscape	Red-cockaded woodpecker	Southeastern Military Bases	Prediction of trends in populations of mobile species across multiple installations

applicable to a wide range of locations and land-use scenarios. This approach can be refined on the basis of needs identified by land managers and the sensitivity of the results to the resolution of available resource information.

HABITAT MODELLING USING SOIL CLASSIFICATIONS AND OTHER SPATIAL INFORMATION IN GEOGRAPHIC INFORMATION SYSTEMS

Management planning for species of conservation concern requires maps of potential habitat, but reliable information from ground surveys is not always available. Geographic information system (GIS)-based habitat models often use vegetation land cover as eco-system analogs, but the location of current vegetation does not always indicate the long-term or potential spatial pattern of habitats. That is, land cover may be more a reflection of recent disturbances than potential conditions. In contrast to current land cover, soil diagnostic characteristics can indicate interactions with climate and land cover for hundreds or thousands of years and are the basis of the soil classification system used by the United States Department of Agriculture National Resources Conservation Service (USDA NRCS) in county-level soil surveys. Therefore, we developed GIS-based habitat models that rely heavily on soil characteristics to predict the spatial distribution of rare habitats. This approach is described in detail by Mann *et al.* (1999).

Example 1: Modelling Habitat of Rare Plants

Background

A GIS model was developed to predict the distribution of potential habitat for rare plant species that occur on limestone geology throughout the eastern United States. At least 19 plant species that are listed by either state or the federal government as threatened, endangered, or of special concern occur in threatened calcareous habitats (referred to as "TCH") that range from rocky or gravelly glades to cedar barrens and surrounding xeric woodland. Four rare species occur in TCH at the Cedar Creek Slope Glades Preserve (referred to as the Preserve) at the Fort Knox Military Reservation, Kentucky. The Preserve is a relatively intact 900 ha tract of TCH. The model was tested for a part of the Preserve and then used to predict occurrences of potential suitable habitat on the rest of the Preserve, including impact areas used for ordnance and tank training.

The Fort Knox Military Reservation occupies 44 150 ha in northeastern Kentucky. Most of the Fort Knox reservation has been surveyed for rare species except for the munitions testing area within the 21 000 ha ordnance impact area. Potential TCH is relatively extensive outside of the Preserve and has been impacted by military vehicle use, logging operations, grazing, and fire.

Approach

Our soil-based model assumes that soils of TCH have developed in response to similar spatial and temporal gradients of edaphic conditions and land cover types (Mann *et al.*, 1999). Based on diagnostic characteristics used in soil classification, we hypothesized that shallow, rocky soils classified as Mollisols are "core" areas of TCH with associated transitional areas of Alfisols. The presence of Mollisols indicates the long-term presence of herbaceous vegetation cover in both forest openings and part of the areas currently occupied by oak red-cedar woodlands.

At a landscape scale, the model assumes that locations of TCH occur at the intersection of data layers containing: (1) soils classified as lithic Mollisols or Alfisols; (2) transitional, barren, urban, or grass land cover; (3) Salem or Harrodsburg limestone geology; and (4) slopes less than 25%. The application of the model was considered successful if the boundaries of TCH openings within the Preserve were contained within or coincident with a predicted patch.

The State Soil Geographic Database and other USDA-NRCS data were also used to predict the distribution of TCH at a regional scale. The map generated from the analysis of the soil and other data was evaluated by comparing it with the county-level distribution of known limestone-glade-endemic species. The success of the model was determined by

XGA 97-0850B/mh

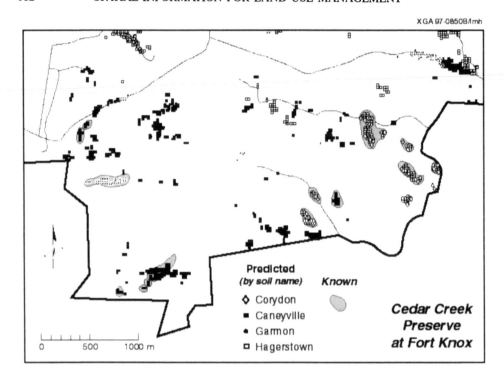

Figure 1.   Predicted and known rare plant habitat locations in the Cedar Creek Preserve at Fort Knox, Kentucky, and the distribution of soils predicted by the soil-based model. Source: Figure 2 in Mann *et al.* (1999).

a chi-square test of co-occurrence of: (1) counties containing predicted calcareous habitat; and (2) counties containing records of threatened and endangered or limestone-endemic species.

Results

The model accurately predicted locations of all ten of the habitat patches that contain TCH within the Preserve at Fort Knox (Figure 1). Discussions with staff at Fort Knox indicated that the predicted potential habitat distribution was accurate, but the model may have over predicted potential habitat in the Impact Zone, where deeper soils are more extensive.

   In the regional application of the model, TCH was predicted in 159 of 562 counties in the eastern United States, and 99 of 158 county occurrences of limestone/dolomite glade-endemic species from the literature were coincident with TCH. Thus, the coincidence of the model with the rare species survey was 63%; expected coincidence was 28% if the distribution of species was independent of predicted habitat within the study area ($\chi^2 = 91$, $p \leq 0.005$).

   The Fort Knox and regional results imply that there is a fundamental similarity in soil genesis and classification in threatened calcareous habitats throughout the region. We found, however, that the soil classification alone was not adequate to identify potentially suitable soil types; information about geology and rock fragments was also needed.

We concluded that current county soil survey data can provide data at a scale appropriate to the occurrence and distribution of TCH patches on the landscape. Although the approach needs further testing in other ecosystems, the use of GIS models incorporating soil taxonomic information should become an increasingly viable approach for predicting the occurrence of rare habitats. In ecosystems in which edaphic constraints are less important than other factors (such as land-use history) additional data coverage would be needed to attain similar accuracy. Disturbance history, such as fire exclusion or grazing, often obscures patterns of edaphically-determined habitat distribution.

Example 2: Habitat Modelling of Lupine Which Serves as a Host for a Rare Butterfly

Background

Fort McCoy is a 24 282-ha training area in the west-central portion of Wisconsin that was established in 1909 as an artillery training area and range. During May through October of every year, Fort McCoy is intensively used as a training site for both Active, Reserve, and National Guard soldiers of the Army, Navy, and Marine Corps.

Fort McCoy is also home to the endangered Karner blue butterfly (*Lycaecides melissa samuelis*). The larvae of the Karner blue are solely dependent upon its host plant, the wild lupine (*Lupinus perennis*). The wild lupine is widely distributed over the training lands of Fort McCoy, and butterflies are abundant. Since the Karner blue became federally listed as endangered in 1992, any military training on Fort McCoy is required by law to minimize its impact on both the butterfly and the lupine.

The modelling approaches that we developed address the issue of how the land managers can meet Fort McCoy's military training needs and yet still ensure that the environmental resources are protected.

Approach

We developed a model for lupine habitat as a surrogate for Karner blue butterfly habitat (Ashwood *et al.* in review). The model adopts an approach similar to that described for the threatened calcareous habitat at Fort Knox using the fact that lupine within the butterfly range is found on sandy soil in post-Pleistocene terrain (Andow *et al.*, 1994). Focusing on soil conditions appropriate for lupine habitat at Fort McCoy, we found three habitat categories as shown in Figure 2: (1) prime lupine/butterfly habitat occurs on west-facing slopes where soils are Boone, Tarr, or Impact; (2) habitat is suitable for lupine where these soil series occur on east facing slopes; (3) lupine does not occur on other soil types or in dense red pine plantations.

Results

Even in locations predicted to be prime habitat, lupine and Karner blue butterfly occurred on only ~13% of the area. GIS models can use physical properties to predict potential habitat, that is, areas that meet criteria believed to encompass the organism's requirements. They cannot predict realized habitat, that is, locations where the organism actually exists. The realized habitat differs from potential habitat because ecological factors such as

Figure 2.   Predicted potential habitat for Karner blue butterfly at Fort McCoy, Wisconsin.

demographics, competition, predation, fragmentation, or barriers prevent the organism from occupying or using even suitable habitat. Thus, the appropriate way to test habitat models is not to monitor for presence of the target organism but to monitor for presence of the habitat itself.

POPULATION MODELS — MANAGING NATURAL RESOURCES AT A SINGLE SITE

Population models provide a means to understand how habitat loss and fragmentation can change the number of individuals in a population over space and time. Changes in spatial patterns of habitat have been implicated in the decline of many bird populations in North

America and elsewhere. Deforestation and the conversion of native grasslands to pasture and cropland reduce the availability of suitable habitat for many species. The increase in edge habitat that accompanies fragmentation can increase brood parasitism and nest predation and can lower the reproductive success of interior species. Habitat isolation can interfere with dispersal and contribute to the decline of local populations. Fragmented landscapes can function as population sinks where reproduction fails to compensate for mortality. Persistence of a species in a sink landscape requires immigration of individuals from more productive source landscapes. Pulliam (1988) introduced the concept of demographic sources and sinks with reference to habitat, but the idea is easily extended to a heterogeneous landscape of multiple habitat types.

Example 1: Spatially-Explicit Population Model of Henslow's Sparrow

Background

Training and testing on DoD managed land can lead to the kind of habitat fragmentation associated with the decline of bird populations. Henslow's sparrow (*Ammodramus henslowii*), for example, is a continental migrant that is of conservation concern at Fort Riley, Kansas, and at Fort Knox, Kentucky. Henslow's sparrow is listed by the Kentucky State Nature Preserves Commission as a species of Special Concern, and the largest documented summer population in Kentucky is found at Fort Knox, where grasslands in the vicinity of Godman Army Airfield are managed as a protected area for the species. Preservation of biodiversity on DoD lands requires assessment of how landscape pattern and changes in landscape pattern affect bird populations. Managers and planners need to know whether the landscapes they manage are sources or sinks for species of conservation concern and whether changes in land use or landscape pattern could shift a landscape from sink to source or vice versa.

Approach

We developed a model of how the spatial distribution of nesting habitat affects the reproductive success of territorial migrant bird species breeding in fragmented, patchy landscapes (King *et al.*, 1999). The model combines a landscape perspective with demographic modelling to investigate how landscape pattern might impact the persistence of avian populations. The landscape perspective derives from a GIS-based model for habitat using the approach described in the previous section. The demographic model uses the information on landscape structure to estimate b_x, the expected number of female fledglings produced by a female of age x. Designed as an assessment tool, the model strives for simplicity and ease of implementation. Accordingly, the model is primarily phenomenological and does not attempt a mechanistic description of avian biology. Model inputs and the data required to test the model are quantities that can be taken from existing literature or might reasonably be collected as part of a demographic study. The model was specifically developed for the assessment of avian demography persistence, and we illustrate its application with an example of Henslow's sparrow at the Fort Knox Military Reservation, Fort Knox, Kentucky. The model has also been applied to Henslow's sparrow at Fort Riley, Kansas. The model's general structure applies, however, to other territorial migrants on other landscapes, both public and private.

Table 2. Results from model simulations of Henslow's sparrow breeding dynamics at Fort Knox, Kentucky (details of model are in King *et al.*, 1999).

Area of potential Henslow's sparrow nesting habitat	859.5 ha
Size of largest patch	51.1 ha
Area of the potential habitat predicted to be utilized	201.4 ha
Number of patches suitable for Henslow's sparrow	3 335
Number (percent) of patches large enough to support nests	21 (0.6%)
Number (percent) of patches smaller than the territory size of 0.4 ha	2 935 (88%)
Number of occupied patches	21
Number of patches responsible for about 80% of fledglings	3
Number of nests supported	100
Number of eggs produced	449 ± 4
Percent of nests successfully fledged	$39 \pm 5\%$
Percent of eggs successfully fledged	$39 \pm 5\%$
Average number of fledglings per pair	1.76 ± 0.21
Number (percent) of the female fledglings	$72 (0.50 \pm 0.03\%)$

In brief, nesting habitat is mapped with a regular grid of square cells. Neighbouring cells are aggregated to form patches. Territories are distributed among patches using incidence functions describing the relationship between species' occurrence and patch area. Nesting success in each patch is a function of the patch's edge to area ratio, reflecting the increased risk of nest predation and brood parasitism associated with increased edge. The number of female fledglings produced in all patches is used to calculate the expected number of female fledglings per female. This demographic variable, an explicit consequence of landscape structure, is combined with survivorship in a life-table model to calculate the demographic indices of net lifetime maternity and the finite rate of increase. These indices provide a simple characterization of the landscape as a population source or sink.

Results

The habitat model predicted only limited breeding success of Henslow's sparrow at Fort Knox (Table 2) (King *et al.*, 1999). The expected net lifetime production of females per female is less than replacement, and the population's rate of increase is less than one. The production of females is insufficient to compensate for mortality reflected in juvenile and adult survivorship, and, in the absence of immigration and assuming constant demographic parameters, the population will decline. The model indicates that the Fort Knox landscape is a population sink for Henslow's sparrow, with an annual rate of decline of approximately 14%. Analysis of the model results suggest that Henslow's sparrow is declining at Fort Knox because of a combination of low reproductive success and low survivorship. Persistence of Henslow's sparrow at Fort Knox appears to require recruitment of individuals from other parts of the species' range. This may represent the historical situation at Fort Knox because the landscape is on the southern edge of the species' summer range and may have always represented "marginal" habitat.

The model we have developed requires further testing at Fort Knox and elsewhere, but we have demonstrated that a simple combination of landscape and demographic modelling can yield a useful assessment tool. Some components of the model need further refinement (*e.g.*, the relationship between nesting success and patch edge:area ratios). However, the quality of the model's results is in large part a function of the quality of the habitat map and biological parameters used as model input. Quality habitat maps are required, and basic life-history data for species of conservation concern are vital. In the absence of such data, we are forced to extrapolate from other species in other regions that may be only weak ecological analogues. The required data can be difficult to obtain and may require long-term monitoring and field work, but they are crucial for accurate assessment of avian demographics and persistence in managed landscapes.

Example 2: Karner Blue Butterfly Population Model

Regional population dynamics occurs in species such as Karner blue butterfly that exist as a set of subpopulations distributed across space that interact with each other by means of dispersal and migration. These patchily distributed and ephemeral habitats suggest that conservation actions for the species need to take into account the spatial arrangement of the butterfly and its habitat. Regional population models offer a means to organize the spatial information and to project spatial patterns of the subpopulations. Therefore, a spatially-explicit population model was developed for Karner blue butterfly and applied to Fort McCoy.

Background

Short-lived adults of Karner blue butterfly emerge in the spring and summer of each year. Eggs are laid on or near wild lupine, on which the larvae obligately feed. The adults obtain flower nectar from several early successional species including the wild lupine. Wild lupine typically occur in a shifting mosaic of patches within open meadows or woodlands that were maintained by wild fires, buffalo wallowing, or other disturbances. European settlers, however, initiated a series of disturbances. Today much of the wild lupine habitat has been lost due changes in disturbance regimes, natural succession, landscape fragmentation, and land-use conversion (Clough, 1992). In Wisconsin, the once 1.6 million ha of savanna are now reduced to 4 000 ha.

Most of the studies concerned with the viability of the Karner blue butterfly have focused on the characteristics of the wild lupine habitat. While the maintenance of endangered species depends on conservation of critical habitats and ecosystems, there is also a need to determine how the butterfly demographics interact with features of the habitat.

Approach

On the basis of the limited data on the population characteristics of the Karner blue butterfly, we are developing a population model that allows us to answer questions about the sensitivity of the Karner blue butterfly to characteristics of the environment and the management actions (Dale *et al.*, unpublished data). The model is designed not only to assess management actions that might impact this endangered species but also to identify characteristics of the butterfly and their habitats for which improved information is needed.

The spatially-explicit, stage-structured stochastic model for Karner blue butterfly is based on habitat suitability maps for Fort McCoy and on demographic data for the butterfly. The habitat suitability maps were developed using a combination of soil classification information, topography and land use similar to that described in the first section of this paper. Relevant demographic data for the model includes dispersal distance for adults; number of individuals per unit area in the egg, larvae, and adult stage; and survival probabilities for each stage of the spring and summer broods.

Results

The spatial distribution of populations is highly sensitive to the dispersal distance of the Karner blue butterfly. With the extremes of dispersal reported from field studies, very different patterns of population distribution result. Thus, the model results illustrate the necessity of including information about dispersal distances in order to understand the regional population characteristics of the Karner blue butterfly.

Nevertheless, there are similarities in model results regardless of the inclusion of different dispersal distances of the butterfly. With largest dispersal distances, the projections indicate fewer patches, but some elements of the spatial configuration remain, such as the patch of wild lupine on the southwestern portion of Fort McCoy, which is a separate population under all scenarios. This pattern suggests the importance of developing a separate management plan for the southwestern population. The presence of a highway that separates that population from other patches of lupine further reinforces this need.

A second set of model runs explored how changes in the characteristics of the population affected the butterfly population structure. By *population structure*, we mean the number of individuals in the egg, larvae, and adult stages over time. The greatest fluctuations in the trajectory were due to changes in survivorship from the egg to larvae stage.

Analysis of the model results indicates that improvements are needed in estimates of the population parameters for adult dispersal and for survivorship from eggs to larvae in order to fully understand the effects of dispersal and survivorship on populations of the butterfly. Clearly these parameters are critical aspects of the butterfly biology. The model results also identify locations of subpopulations of the butterfly at Fort McCoy that require particular conservation focus. Thus the Karner blue butterfly model illustrates two benefits of using landscape population models in efforts to enhance conservation: they can identify parameters for which uncertainties need to be reduced, and they can indicate features of the spatially-defined subpopulations that require management attention.

TRANSITION MATRIX MODEL — PREDICTING THE RESULTS OF LAND MANAGEMENT ACTIONS

Example 1: Land Uses at Fort McCoy

Background

Application of land-use principles requires the use of spatial data and tools to analyze those data. Modelling tools have been developed as a way to relate these data to potential land-use decisions. For example, land-use models typically project land-use changes using maps and consider their spatial implications, such as changes in edge effects or habitat fragmentation. One of the tools that proves useful for assessing the effects of decisions on ecological properties is the transition matrix model. These transition matrices simulate the probability of a change from one successional state or cover type to another and to project changes over time and space in land cover and land use. Landscape transition matrices offer a way to consider the ecological impacts of future land-use activities. We have developed and applied the transition approach in such a way that ecological impacts on a system can be included in the decision-making process (Dale *et al.*, in press).

Approach

Landscape transition matrices simulate temporal and spatial changes in land use and land cover. A landscape transition model can be used to map and assess the impact of land-use activities on natural and cultural resources. Land-use activities can be characterized by using a common set of parameters (magnitude, frequency, areal extent, spatial distribution, and predictability) that can be applied either to specific activities or to different intensities of the same activity. This approach permits evaluation of the incremental and cumulative effects of diverse activities, such as road building, military maneuvers, grazing, timber harvests, or environmental restoration. Evaluating the risk posed to habitats and species can be expressed as the probability of a decline or enhancement in the abundance of guilds, species, or their habitat. Such an approach is generic, and with appropriate databases it can be applied to any site.

Results

The transition approach was applied to land uses at Fort McCoy (Dale *et al.*, in press). As an example, the ecological risk map for tracked and wheeled vehicle training shows that more than half of the area with pines, oak, and grass/rock/brush cover types to be at risk of change (Plate 18). These cover types are distributed throughout the Fort McCoy area.

The training impacts would increase the number of patches of these cover types and reduce the sizes of the largest patch and the average patch. In addition, there may be secondary impacts on the species that use these cover types as habitat. According to the lupine impact map (Plate 18), the risks that wheeled and tracked vehicle training poses to lupine are not uniformly spread across the installation. Overall, 56% of the lupine sites would be at risk. This amount of risk justifies an active management program for these lupine sites.

In addition to use in managing natural resources, the landscape approach is directly applicable to: (1) planning for facility closures and realignment (*e.g.*, identification of facility closures that provide the best conservation opportunities); (2) developing environmental restoration and waste management strategies; (3) supporting compliance with the Endangered Species Act, the National Historic Preservation Act, the National Environmental Policy Act, and the Executive Orders for Floodplains and Wetlands; and (4) developing integrated risk assessments that address cumulative effects.

REGIONAL MODELS — MANAGING NATURAL RESOURCES ACROSS SITES

Example 1: Red-Cockaded Woodpecker Model for Southeastern United States

Often it is important to understand population dynamics of species across a large area. Regional population dynamics models offer tools by which to explore large-scale dynamics. This need arises in examination of the red-cockaded woodpecker (RCW; *Picoides borealis*), a federally listed endangered species endemic to mature pine forests of the southeastern United States. Populations of RCW exist on many DoD installations across the Southeast, and their occurrence impacts the DoD training and testing mission.

Background

The preferred pine-bunch grass savannah habitat of the red-cockaded woodpecker was once widely distributed across the southeastern United States. However, fire management, deforestation, forestry practice, and other changes in land use have reduced and fragmented this once common and relatively contiguous ecosystem type. The distribution of longleaf pine (*Pinus palustris*) alone has been reduced from perhaps 37 million ha prior to European settlement to currently less than 1.2 million ha. Consequently, remaining populations of RCW are fragmented, small, and isolated, persisting primarily on federal and state properties and closely associated private lands. The U.S. Fish and Wildlife Service recovery plan for RCW calls for the establishment of at least 15 viable populations across the species' range. DoD lands figure prominently in the recovery plan; 6 of the 15 proposed recovery populations involve DoD installations. It is important to understand the role of DoD installations within the context of regional RCW recovery and management.

Approach

The fragmented and isolated occurrence of RCW populations is characteristic of a regional metapopulation, and regional metapopulation dynamics might allow for the persistence of the species despite the highly fragmented distribution of pine-bunch grass habitat. It is not clear, however, that the 15+ populations of the USFWS meet all the criteria of a metapopulation. For example, because of the relatively recent history of RCW habitat fragmentation and insularization, an *a priori* assumption of a regional metapopulation is unwarranted. Therefore, we do not assume a metapopulation structure for the red-cockaded woodpecker. Rather, we ask a series of questions:

1) Can the red-cockaded woodpecker be managed as a regional metapopulation?
2) Is the current distribution of red-cockaded woodpecker habitat consistent with long-term regional persistence?
3) If not, what changes in habitat distribution or other management interventions would promote regional persistence?

Similarly, we do not assume a metapopulation model but instead adopt a patch-based modelling approach from which metapopulation dynamics may or may not emerge. Our patch-based model of change in RCW clusters takes the following form:

$$dN_i/dt = r_i N_i (1 - N_i/K_i) - D_i + C_i \qquad (1)$$

where

N_i = number of active RCW clusters in population i;
r_i = within-population rate of change in active clusters;

K_i = local carrying capacity (number of active clusters) of the population;
D_i = disturbance function specific to population i;
C_i = colonization of population i from other populations.

D_i is due to environmental stochasticity (*e.g.*, hurricane damage) and $C_i = \sum_j c_{ij}$, where c_{ij}, the colonization of population i from population j is a function of the distance from population j and the number of active clusters in population j. Details of the model can be found in King *et al.* (in review).

We have parameterized this model with information currently available for red-cockaded woodpecker in the southeastern United States (*e.g.*, distances between clusters). We also included the infrequent long-distance dispersal event that can connect otherwise disconnected populations in subtle but significant ways. We have examined records of long-distance dispersal for RCW across its range and used this information to calibrate the model's colonization function.

Results

Mean within-population dispersal of RCW juveniles is on the order of 5–10 km. Dispersal distances of up to 30 km are not uncommon, however, and dispersal by juveniles of 160 and 287 km have been reported. We use these dispersal records to characterize connectivity among populations of RCW on federal lands of the southeastern United States. For example, the population at Fort Stewart, Georgia is isolated from all other populations at dispersal distances of less than 100 km, it is connected to four populations at dispersal distances of 200 km, and it is connected to 12 populations at dispersal distances of 300 km. Similarly, the RCW population at Fort Benning, Georgia, is isolated from all other populations at dispersal distances up to 100 km, connected to four populations at dispersal distances of 200 km, and connected to 10 populations at dispersal distances of 300 km.

Long-distance dispersal connecting otherwise isolated populations can increase the potential of regional RCW recovery and persistence. In Plate 19, we show results from the model applied to the 15 designated recovery populations without (top) connectivity by long-distance dispersal and with (bottom) connectivity by long-distance dispersal. Connectivity by means of long-distance dispersal promotes colonization which increases the rate at which the recovered populations reach their capacity for active RCW clusters.

CONCLUSIONS

To be useful for preserving the natural diversity and habitat on military installations, each of the ecological models is designed; (1) to use information that is available at most installations; (2) to be adaptable to other similar concerns; and (3) to have products that reflect the concerns of land managers. These are the characteristics that make models generally applicable to both site-specific and regional decisions. Such models are useful to military land managers as well as other private and public land managers.

The benefits from these ecological models are fourfold. First, a method to identify locations of habitat and to characterize key attributes of species at risk is set forth. Second, a framework to analyze potential impacts of land-use activities on natural resources is developed. Third, case studies demonstrate these approaches at locations at Fort Knox, Fort McCoy, and the southeastern military installations. Finally, the approach links management questions with models.

ACKNOWLEDGMENTS

Comments on the paper by Robert O'Neill and Dan Jones were quite helpful. Dave Aslesen and Tim Wilder from the Fort McCoy Military Reservation provided data and discussed installation's management concerns with us. Cathy Carnes graciously provided unpublished reports and shared her experiences dealing with the Karner blue butterflies and wild lupines. Raymond McCord assisted in coordination with the military installations. Desmond Fortes contributed to the transition modelling study. Bill Hargrove and Robert Washington-Allen assisted with some of the spatial analysis and mapping.

The project was funded by a contract from the Conservation Program of the Strategic Environmental Research and Development Program (SERDP) through the U.S. Department of Defense Military Interagency Purchase Requisition No. W74RDV53549127 and the Office of Biological and Environmental Research, U.S. Department of Energy (DOE), under contract DE-AC05-96OR22464. Oak Ridge National Laboratory is managed by Lockheed Martin Energy Research Corporation for DOE under contract DE-AC05-96OR22464. The paper is Environmental Sciences Division Publication No. 4911.

REFERENCES

1. D.A. Andow, R.J. Baker and C.P. Lane (editors) (1994) *Karner Blue Butterfly: A Symbol of a Vanishing Landscape*, Minnesota. Agricultural Experiment Station, Miscellaneous Publication 84-1994. (University of Minnesota, St. Paul).
2. T.L. Ashwood, L.K. Mann, A.W. King, V.H. Dale and W.W. Hargrove *Natural Areas J.*, (in review).
3. M.W. Clough (1992) *Fed. Reg.* **57**, 59236.
4. V.H. Dale, A.W. King, L.K. Mann, R.A. (1998) Washington-Allen and R.A. McCord, *Environ. Manage.*, **22**, 203.
5. V.H. Dale, D.T. Fortes and T.L. Ashwood, A landscape transition matrix approach for land management, in *Integrating Landscape Ecology into Natural Resource Management*, edited by J. Liu, (in press).
6. A.W. King, L.K. Mann, W.W. Hargrove, T.L. Ashwood and V.H. Dale (1999) *Assessing the persistence of an avian population in a managed landscape: A case study with Henslow's Sparrow at Fort Knox, Kentucky*. (Oak Ridge, Tenn.: ORNL/TM 13734. Oak Ridge National Laboratory).
7. M. Leslie, G.K. Meffe, J.L. Hardesty and D.L. Adams (1996) *Conserving biodiversity on military lands: a handbook for natural resource managers*. (The Department of Defense Biodiversity Initiative, Office of Deputy Undersecretary of Defense, Environmental Security, Washington, DC).
8. L.K. Mann, A.W. King, V.H. Dale, W.W. Hargrove, R. Washington-Allen, L. Pounds and T.A. Ashwood (1999) *Ecosystems*, **2**, 524–538.
9. H.R. Pulliam (1988) *Amer. Nat.*, **132**, 652–661.

13. PREDICTING LAND USE CHANGE IN AND AROUND A RURAL COMMUNITY

Bruce Maxwell, Jerry Johnson and Clifford Montagne

INTRODUCTION

Many counties in the Rocky Mountain West of the United States are experiencing a population surge that promises to surpass the impact of the first western migrations (Rudzitis, *et al.*, 1996). Seven of the ten fastest-growing states in the nation are in the Rocky Mountain West (U.S. Bureau of the Census, 1995) and 67% of the counties in the Rocky Mountain axis grew at rates faster than the national average (Nelson and Beyers, 1998). In Montana, annual population growth rates in several counties have reached double digits since 1990. For example, Ravalli County experienced 30% population growth between 1990 and 1995. Gallatin and Flathead counties reported almost 18% growth over the same period. This rapid population growth results in significant land use change in and around many small rural communities. These landscape changes have ecosystem and socioeconomic impacts not clearly understood prior to change occurring although this is when land use planning and policy could minimize negative impacts of growth in the community. However, prediction of land use change can be a valuable tool for rural communities to plan for the future. Spatially referenced information is essential for accurately predicting land use change and subsequent ecosystem and socioeconomic impacts in these communities.

Predicting land use change in areas with a mosaic of private ownership requires combination of a wide range of driving variables. Land capability can be determined by soil, climate and land cover variables that are often available in geo-referenced form (Dale *et al.*, 1993; Berry *et al.*, 1996; Turner *et al.*, 1996). However, socioeconomic variables that determine land use change under private ownership have not been thoroughly explored (Wear and Flamm, 1993). We summarize a range of variables that may be considered drivers of land use change in rural communities of the western U.S., and use a case study to demonstrate a method for capturing the influence of these diverse variables to predict land use change. We also describe an ecosystem integrity index that summarizes the biological impacts of land use change, and discuss some of the socioeconomic implications associated with land use change.

POPULATION GROWTH AND LAND USE CHANGE IN THE ROCKY MOUNTAIN WEST

Explanations for rural population growth are complex, but can be summarized in the quality of life model *vs.* the demand-driven model (Johnson and Rasker, 1995). The quality of life model suggests that choice to move to a rural area is a function of a mix of amenities

173

acting as pull factors. Pull factors include small towns, low crime rates, a desirable climate, recreation opportunities, or good quality jobs (Shepard, 1993; Power, 1995). The demand-driven model argues that immigration to rural areas is a function of wages and employment. This model of rural migration suggests that increases in the demand for labour create migration to the community (Lowry, 1966; Fabricant, 1970). An example is migration to a rural town to work in a resource extraction industry. This model also suggests that periodic "disturbances" encourage growth, typical of "boom" cycles in extractive sectors. Such growth tends to be short term (oil and gas field development, international grain sales, *etc.*). The driving variables that predict land use change in these models are difficult to use since they often arise away from the geographic region of interest; a mine may be 50 to 100 km away from the community where workers reside although it subsequently impacts the landscape and socioeconomic structure of that community.

Other explanations focus on structural changes in urban and rural systems. The large number of retirees in the U.S. enables a significant proportion of the population to live in almost any location they find desirable. Sunbelt migration is well-documented, but others prefer mountain resort locations for their retirement. Another explanation is the globalization of the national economy which has facilitated a regional transformation of place of residence and work. Advances in telecommunications and transportation technologies allow entrepreneurs and telecommuters to move to rural areas and work from homes and small offices (Nelson and Beyers, 1998). As populations in rural areas increase, service related employment follows. Personal and other support services such as health care, investment counseling, transportation, and a host of other services are required and provide employment opportunities. This booming service economy is accompanied by a healthy tourism economy. In the seven states that comprise the Rocky Mountain core,* the share of total employment attributed to tourism is approximately 6.5% (381 300 jobs) (Rafool and Loyacono, 1997). Thus, there are diverse factors influenced by national and global scale population trends that influence land use change and subsequent impacts in rural communities.

This rural land use transformation is made possible by the aging of the agricultural landowner population. As agricultural landowners enter retirement age much of the land in production is made available for development. Farms are consolidated by corporate interests, children of agricultural producers express less desire to enter agriculture, and inheritance taxes create difficulties for large landowners that seek to pass land on to family members. A frequent solution for agriculture producers is to subdivide the land and sell to rural immigrants. Associating landowner age with land use change, especially urbanization (subdivision development), may be a suitable relationship to explore for predicting land use change.

The effects of rural land use change are complex and, with respect to residential growth, are twofold; social effects on rural communities and ecological effects of rural development. The notion of rural western communities is of a close-knit social and economic structure with employment, social, and consumer functions carried out in a self-sufficient socioeconomic system (Carroll and Lee, 1990; Kemmis, 1990; Jobes, 1988). Some however, have observed that recently arrived rural residents are less attached to the nearby

---

*The Rocky Mountain states include: Idaho, Montana, Wyoming, Utah, Colorado, New Mexico and Arizona.

Table 1. Socioeconomic and ecosystem impacts of land use change.

Social & Community Effects of Rural Land Use Change	
Changes in landowner structure	Turner, Wear and Flamm, 1996
Changes To Community History & Culture	Williams and Jobes, 1990; Jobes, 1988; Rudzitis, 1996; Beggs, Hurlbert and Haines, 1996
Impact On Agricultural Lands	Greene and Harlin, 1995; Heimlich and Vesterby, 1992; Archer and Lonsdale, 1997
Impact On Open Space/View	Gersh, 1996; Nieman Reports, 1996
Uneven Cost Of Residential Service	Haggerty, 1996; American Farmland Trust, 1992; Kelsey, 1998
Changing Political/Economic Structure	Alma and Witt, 1997; Beyers, Lindaho and Hamill, 1995
Quality Of Life Effects	Johnson and Rasker, 1995; Jobes, 1988; Decker and Crompton, 1993
Ecological Effects of Rural Land Use Change	
Water Pollution & Sewage	Gersh, 1996; LaGro and DeGloria, 1992
Fragmented Habitat	Theobald, 1998
Threats To Biodiversity	Pimental *et al.*, 1992; Farrier, 1995; White *et al.*, 1997; Forester and Machlis, 1996
Land Use Conversion	Riesame, Gosnell and Theobald, 1996; Bean and Wilcove, 1997

community and have less need of it to meet social and economic needs. In the new rural west, rural land adjacent to small towns serves primarily as residential rather than social locations (Johansen and Fuguitt, 1990). The effect is that main street shopping functions move to regional shopping centers and small town business districts erode. In the process, the social function of downtown also erodes (Snepenger, Reiman, Johnson, Snepenger, 1998). The community also loses its sense of place and solidarity (Huang and Stewart, 1996).

Increased development pressure also results in significant impacts to ecological quality. These effects include threats to habitat (*i.e.* fragmentation, loss), geographic features (i.e. water supply and quality, soils), and native species (*i.e.* weed invasion, biodiversity). The conversion of native and agricultural lands to residential subdivisions or small ranches is of particular concern because such development will probably never revert to undeveloped land (Riebsame *et al.*, 1996). Many of the social and community effects, as well as the ecosystem affects of rural growth have been summarized in the literature (Table 1).

CASE STUDY

We studied a community in detail to demonstrate the use of spatially-referenced socio-economic information in combination with standard geographic variables to predict land use change. The spatial context of the information allowed physical and biological

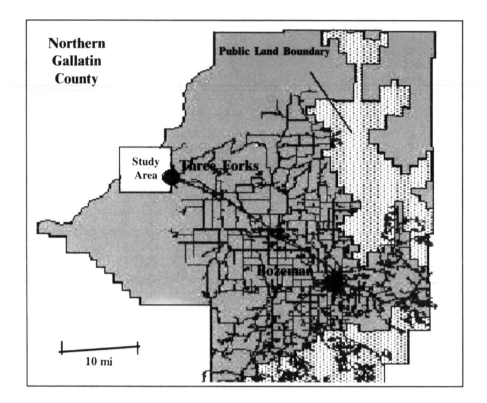

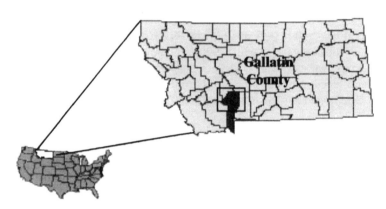

Figure 1. Three Forks, Montana and surrounding area "school shed" was used for a case study.

landscape features to be linked with socio-economic variables to change behaviour and attitudes in the community. The study community, Three Forks, and surrounding land are in Gallatin County, Montana, USA (Figure 1). The High School District boundary ("schoolshed") was used as the study area boundary because the school is the focus of the social fabric of the community. Employment is in industrial and agricultural sectors as well as in markets for agricultural and ranching activities. Community involvement in social events such as rodeo and school sports, health care planning, and land and water issues are closely linked. Historically the social and economic system is based in an extraction economy, the agricultural history of the area, its geographical location near an interstate highway, and proximity to a rapidly growing urban center (Bozeman, MT. pop. 35 000); this creates potential for many of the real and imagined social tensions typical of communities in the rural west.

A thorough description of the socioeconomic background of the community is crucial to develop variables that are associated with land use change. The Three Forks area economy is historically based on railroad employment, agriculture (dry-land and irrigated wheat), mining-related manufacturing (cement, talc products, gold), a small lumber mill, and a small retail sector. The mining industry is economically stagnant. Recently, a vertically integrated bakery has become established which has provided a market for locally produced wheat and provides value-added products to local agriculture. Some small businesses have appeared, including a fishing lodge/motel, a small cement products plant, a B&B/restaurant, and a wholesale tool supplier.

The population of the study area has increased in recent years, even with few new employers. House prices increased by 25% between 1992 and 1996 due to pressures from Bozeman, 70 km to the east. Employment in home construction increased following transformation of the railroad right-of-way into a housing subdivision. Homes in this and other subdivisions were purchased as soon as they became available. Older homes in the region are being acquired and refurbished.

Many residents commute to Bozeman which provides diverse employment opportunities. In addition to the structural employment created by a regional population of almost 50,000, several relatively large employers in the Bozeman area provide jobs in high technology, tourism, consumer services, light manufacturing, and education. As a result of increased employment opportunities in Bozeman, more Three Forks residents also shop and enjoy entertainment activities in Bozeman and this has displaced several Three Forks businesses.

CASE STUDY OBJECTIVES

The Three Forks project is a study of a rural community in transition from one based primarily on agriculture and natural resource extraction to one based on a service economy. Three Forks is in the process of acquiring the social and economic traits that typify changes taking place across the rural west. The objective is to assess and predict the rate, nature, and impact of land use change in and around the community. A second objective is to develop modelling tools that meet the criteria of being user-friendly (for community employees, volunteers, high school students, etc.), relatively low cost, and applicable to rural communities.

CASE STUDY METHODS AND TOOLS

A series of meetings were conducted with community leaders who expressed the desire for visual tools that demonstrate growth and changes in the community, and that can be used for planning purposes. Landscape planning in the West is often a tenuous process

Table 2. Land use/cover class categories identified on aerial photos.

1. Unclassified (not on map)
2. Residential
3. Industrial/Commercial
4. Transportation (highway right-of-way)
5. Water (ponds, lakes, rivers)
6. Upland shrub/grass rangeland
7. Dryland crop production (wheat; barley; oats; etc.)
8. Dryland hay production
9. Conservation Reserve Program (planted perennial grass to replace crops)
10. Irrigated crop production (wheat; barley; oats; etc.)
11. Bottom-land native range (shrubs and grass lands in 100 yr flood plain)
12. Bottom land hay production
13. Riparian (river-bottom cottonwood and willow stands)

marked by lack of public and political support because of perceived threats to private property rights and competing political agendas. However, meeting participants considered visualization of future landscapes through predictive modelling to be a powerful tool for public assessment of growth alternatives as it gives residents the opportunity to anticipate future landscapes. In addition, planners at city and county levels can use growth predictions to estimate future requirements for public budgets, infrastructure enhancements and public services. A GIS is the logical tool to depict present conditions and provide the community with future growth scenarios.

Land use was classified (Table 2) within the 60 km^2 study area on a 2.5 ha resolution grid from 1:24,000 aerial photos taken in 1965, 1979, 1984, 1990, 1994 and 1995 (Table 3). This historic land use information was stored in ArcView GIS and used to generate a 2+n dimensional probability transition matrix for land use change between any two dates. Additional information in the form of maps (Plate 20) were also created for use as factors influencing transition probabilities in the matrix including:

Table 3. GIS layers of information used for predicting land use change over time near Three Forks, MT.

Data Type	1965	1979	1984	1990	1994	1995	1998
Land Use	✔	✔	✔	✔	✔	✔	✔
Roads	✔	✔	✔	✔		✔	
Streams				✔			
Distance from streams	✔	✔	✔	✔	✔	✔	✔
Distance from roads	✔	✔	✔	✔	✔	✔	✔
Nearest Neighbor	✔	✔	✔	✔	✔	✔	✔
Soil					✔		
Landowner Behavior					✔		

1) Soil type;
2) Nearest neighbor cell type;
3) Distance to streams (Plate 20 top right);
4) Landowner land use change behaviour (Plate 20 bottom left); and
5) Distance to roads for each year of land use classification (Plate 20 bottom right).

The Land Use/Land Cover Change Prediction System

The Land Use/Land Cover Change Prediction System (LUCCPS) was designed to create a visualization (map) of future land use based on historic changes and be a tool for community planning. Predicted maps are produced, digital pictures of the landscape are linked to the map at familiar viewpoints, and digital photos are constructed to match previous year land use maps by removing structures and roads. Similarly, photos were edited to show predicted landscapes based on the predicted maps, by adding structures and roads to provide a vivid impression of different land use changes and projections.

The LUCCPS model is based on the history of past land use change, natural features, man-made infrastructure, and land use decisions. Our approach to predicting change concentrates on identifying and incorporating independent driving variables into the transition probabilities for landscapes in private ownership. Our interest is in capturing a full compliment of the independent driving variables to reflect the wide range of land management objectives that maybe present.

LUCCPS allows users to select one or more data layers. Data from any or all of the layers can be used to compute a transition matrix. A minimum of two years of land use/cover data must be selected. The first land use/cover data layer is the *primary* layer, and the second is the *response* layer. The primary and response layers are typically from sequential years, reflecting historical change. Additional data from any other layers can also be selected as *secondary influences*. LUCCPS currently allows up to thirty secondary influencing layers.

Once all of the data sources are selected, LUCCPS calculates a transition matrix. This matrix represents the probability that a combination of category values in the same grid cell in the primary and the secondary layers corresponds to a specific category in the same grid cell in the response layer. The dimensionality of this matrix is equal to the number of secondary influences plus one for the primary and one for the response.

Consider a two-dimensional matrix computed from one primary and one response layer, with no secondary influences represented by the following two grids (with X and Y coordinates) for two years (1980 and 1985). The numbers represent different land use/cover categories:

Primary (P) Layer
1980

Y\X	X1	X2	X3
Y1	1	1	3
Y2	1	2	2

Response (R) Layer
1985

Y\X	X1	X2	X3
Y1	1	1	3
Y2	2	2	2

If 1980 is chosen as the primary layer and 1985 as the response layer, the following is the transition matrix:

P\R	1	2	3
1	.66	.33	0
2	0	1	0
3	0	0	1

The transition matrix shows that two of three times, a category **1** in the primary year remains **1** in the response year, while one of three times it changes to **2**. It never changes to a **3**. Values of **2** and **3** stay the same between the primary and response years. Note that each row in the matrix must sum to one, and each value in the primary layer must have a value in the response layer. Now consider a secondary influencing data layer for same area but a different set of categories. For example,

		1980	
Y\X	X1	X2	X3
Y1	A	B	A
Y2	B	C	A

These data are a secondary influence and the transition matrix now has three dimensions, represented using a table with a column for the index value in each dimension, plus a column for the contents of the specified entry in the matrix:

	R=1				R=2				R=3		
P\S1	A	B	C	P/S1	A	B	C	P/S1	A	B	C
1	1	0.5	0	1	0	.5	0	1	0	0	0
2	0	0	0	2	1	0	1	2	0	0	0
3	0	0	0	3	0	0	0	3	1	0	0

Again, the probability for all entries with the same primary and secondary influence values must either sum to zero (indicating that the combination was never observed) or one. The first value of one in this matrix indicates the combination of **1** in the primary and **A** in the secondary always results in a **1** in the response. However, a **1** in the primary and a **B** in the secondary result in either a **1** or a **2** in the response layer, each result having likelihood 0.5.

The transition matrix is populated by traversing the map and identifying the combinations of categories for the input layers. Each entry in the matrix represents the number of times the corresponding combination of category values is observed. To convert these counts to probabilities, each entry is divided by the total number of times that its combination of primary and secondary influence values occurred.

After the probability transition matrix is computed, it is used to predict land use/cover in future years (see below). The grid is again traversed, and the land use/cover class for each cell determined as a predicted class using a uniform random number generator to

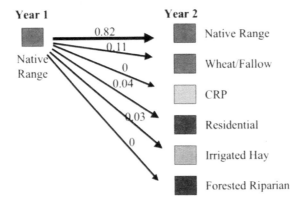

select a class based on probabilities in the transition matrix. This retains the simple deterministic character of the prediction system; the system does not use heuristics or hard-coded decision points to make predictions. Different scenarios of land use change are created in LUCCPS by choosing different primary, response, and secondary influencing data layers.

Ecosystem Integrity Index

Field observations in 1995 and 1996 were used to develop an ecosystem integrity (EI) index for each land cover/use type. The parcels of land measured were those consistently in specific land uses since 1965. An EI index was assigned to each grid cell on historic as well as predicted maps. EI values are accumulated for study area for a given year to give a general trend in ecosystem integrity associated with different land use change scenarios.

The EI is composed of 14 variables for upland and 13 variables for valley bottom areas (Table 4), reduced from an initial list of 32 vegetation, animal and soil variables. Variables were reduced using stepwise multiple linear regression. A variable that does not significantly contribute to the model is dropped. Sample plots were ranked according to their

Table 4. Variables used to construct the upland and river bottom land ecosystem integrity index.

Upland			
bare soil	water infiltration rate	soil organic matter	soil nitrate/nitrogen
moss cover	soil bulk density	soil carbon/nitrogen	soil nitrate
litter cover	plant morphological diversity	plant species richness	canopy cover
root biomass	plant canopy diversity		

River bottom			
% weeds	soil bulk density	water infiltration rate	plant canopy diversity
% litter	total soil nitrogen	plant species richness	plant morphological diversity
% canopy cover	% downed wood	soil carbon/nitrogen	% rock or pavement
root biomass			

raw EI score and referenced to the sample plot with the highest score, to create a relative score.

Socio-economic Variables

Spatially-referenced socio-economic variables were developed from a series of meetings with community leaders, a survey of attitudes about land use (Johnson and Maxwell, 1996), and a set of interviews with local land owners. The interviews were qualitative; they were semi-structured and aimed to provide insights into the behaviour of the respondents. Their goal was to explore the range of landowner decision making. Most of the information collected identified factors important to the community from which to formulate a set of socio-economic variables that influence changes in the landscape. The land use change behaviour survey grew from the observation that a few landowners have control of large portions of the landscape. By asking these landowners why they made land use change decisions that we had observed on the aerial photos, we were able to categorize their behaviour into 6 general categories:

1) Economic return motivation;
2) Tradition motivation;
3) Government policy motivation (conversion to Conservation Reserve);
4) Economic return and conservation;
5) Economic return and community service (providing new areas for housing);
6) Family considerations.

These categories are associated with each owner's land or portion of land with a particular biogeographic classification. For example, some owners are willing to place riparian zones in Conservation Easements, but retain dry upland areas for crops based on maximizing net economic returns.

CASE STUDY RESULTS

Land use changes that were most dramatic in the study area were precipitated by government policy through the Conservation Reserve Program (CRP). Approximately one third of cropland was converted to CRP between 1985 and 1988 (Plate 21). This had a large impact on the landscape during the years of our study. CRP enlistment was contemporaneous with the loss of some farm machinery businesses, which may have been a result of large tracts of land no longer being actively farmed. Land enlisted in CRP requires a 10 year commitment to receive funding; farm land that may have been subdivided for housing was maintained as open space under CRP reducing urbanization rates. CRP can increase stability in farm income which may also change consumer behaviour.

Each location was assigned an EI score corresponding to the average EI score for the land use represented at that location. EI scores for each location for each year were summed. The range of scores was similar between the valley bottom and upland areas and these areas were grouped together to develop the study area EI. In 1965, the EI of the study area was 1064. The relative proportions of land in dry land wheat production and

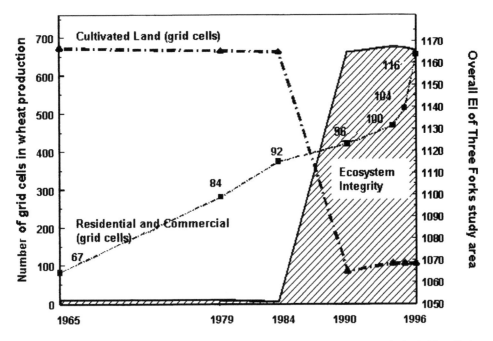

Figure 2.   The relationship between wheat production, urbanization and ecosystem integrity in the Three Forks, MT study area.

range in 1965 were the major influences on EI. The EI remained at the same level (1072) in 1979. The introduction of CRP between 1984 and 1990 caused an increase in EI from 1070 to 1184. In 1994 the overall EI was 1185 as irrigated alfalfa was rotated to irrigated wheat. The study area EI was 1183 in 1995 and stayed the same in 1996 despite the conversion of a small area of CRP to a residential area. CRP had the greatest impact on ecosystem integrity in the study area (Figure 2). Ecosystem integrity was not significantly influenced by the constant increase in residential area. This is in part due to the large area converted to CRP while most of the residential increase was along the railroad right-of-way. This was a conversion from industrial classification to residential, mean EI being higher in residential than industrial areas.

To accommodate socio-economic or ecological change in rural communities, public policy can be designed to mitigate, encourage, or otherwise manage land use change. Community support for land use policy can be gained through collaborative "visioning" of growth scenarios. Close geographical ties to the rural landscape make GIS a particularly effective tool for both envisioning change and planning future appearance by a community.

Using LUCCPS we demonstrated the potential impact that CRP was having on urbanization in the Three Forks area. We predicted growth under two different scenarios. In the first, the transition matrix was based on years prior to CRP; in the second from years during CRP. The area of residential development predicted was higher using the pre-CRP matrix (Figure 3). Three Forks community leaders found these results somewhat surprising as the connections between CRP and community growth are not intuitive.

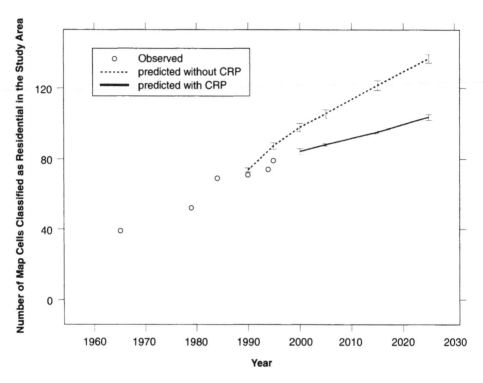

Figure 3.   Observed number of map cells classified as residential and predicted number of residential map cells under two different policy scenarios: 1) "before CRP" and 2) "during CRP".

Results from landowner behaviour interviews indicate that agricultural landowners operate on a continuum from profit maximization to value-driven decision making (Oglethorpe, 1995). Dent *et al.* (1995) describe the management of farm systems as a complex mix of elements contained within two broad categories. The first includes economic factors such as profit maximizing behaviour, commodity prices, relative value of land and product mix. The second includes social factors such as personal values, lifestyle pursuits, family considerations, and past experience. The mix of these elements is included in our model through a land use behaviour index used as a dimension in the probability transition matrices. Including the behaviour index significantly improves predictions, indicating the vital connection between social factors and land use change.

The final objective is to convert the results of LUCCPS predictions into visual representations of predicted landscapes. Specifically, we provide digital images that community residents, planners, and local government officials can use in several ways. First, they help to integrate and visualize impacts on landscapes, viewsheds, ecosystem integrity, and other variables considered important by the community. Second, they help to determine if possible impacts are desirable. Third, images may be used as a basis for entering into dialogue with land managers before a land management decision takes place. The series of images (Plate 22) show the original undeveloped agricultural landscape. By year 2005 the LUCCPS model indicates that the land will move into a rural residential development phase. Based on other development in the area we represent the growth to appear as a non-clustered 2–3 acre subdivision with little to no agricultural function and significant development on the forest edge. This scenario is placed before area residents as the outcome most likely for the current development path and provides them and policy makers with a choice over the future of the regional landscape. Feedback from the exercise includes enhanced community discussion of growth preference; ecosystem integrity measures can be associated with differing land use change alternatives for additional evaluation.

The four objectives of tool design were met and all products of the research are being integrated into LUCCPS. The design of the system is such that drivers of land use change can be merged to provide greater predictive capacity. Additionally, scenarios can be run that provide insight into future policy changes and resulting effects.

The concept of applying ecosystem management principles to agriculture appears to be validated by our results. Diversity of agricultural production mixed with land in a semi-natural state benefits the economic stability of the community and we provide some measures and indicators of different aspects of these land use impacts for land use change scenarios. Comparison on indicators is necessary. For example, EI could be weighed against income potential and satisfaction with quality of life to determine the community structure and land use pattern that most favors community sustainability. The EI index may also prove useful for planning to minimize adverse impact to the landscape. Analysis of landscape preferences based on photographs may play a useful role in establishing the balance sought by a community, developers and agricultural producers and provides an opportunity to develop LUCCPS further.

CONCLUSIONS

The study area was selected based on its proximity to a small community and the diversity of land uses representative of the Rocky Mountain Region. In addition the community exhibits the main socio-economic characteristics of the Rocky Mountain West. Accurate land use prediction in different geographical locations will require the collection of different sets of data layers although the framework and methodology of LUCCPS is transportable. Similarly, land uses in different environments will have different relative EI rankings. Regional differences in farming practices may also contribute to differences in EI for different land uses. The methodology for assessing EI and quantifying land use behaviour in rural communities coupled with the ability to predict future land use has application throughout the Inter-mountain West of the United States. Our land use prediction system, coupled with the EI index, may be useful for land use planners concerned with maintaining the overall social and ecological integrity of rural communities. The ability to produce visualization of future land use scenarios is also of value to communities concerned with the effects of population growth and associated residential development. Clearly, the pictures provide an important visual impression of different scenarios of landscape change.

REFERENCES

1.   L.R. Alm and S.L. Witt (1997) *Soc. Sci. J.*, **34**, 271–284.
2.   J.C. Archer and R.E. Lonsdale (1997) *J. Rural. Stud.*, **13**, 399.

3. M.J. Bean and D.S. Wilcove (1997) *Conserv. Biol.*, **11**, 1–2.
4. J.J. Beggs, J.S. Hurlber and V.A. Haines (1996) *Rural Sociol.* **61**, 407–426.
5. M. Berry, R.O. Flamm, B.C. Hazen and R.M. MacIntyre (1996) *IEEE Computat. Sci. Eng.*, **3**, 24–35.
6. M.S. Carroll and R.G. Lee (1990) Occupational Community and Identity Among Pacific Northwestern Loggers: Implications for Adapting to Economic Changes, in *Community and Forestry: Continutities in the Sociology of Natural Resources*, edited by R.G. Lee, D. Field and W.R. Burch. (Boulder CO: Westview Press).
7. V.H. Dale, V. O'Neill, M. Pedlowski and F. Southworth (1993) *Photogramm. Eng. Remote Sens.*, **59**, 997–1005.
8. J.M. Decker and J.L. Crompton (1993) *J. Profess. Services Market.*, **9**, 69–94.
9. J.B. Dent, G. Edwards-Jones and M.J. McGregor (1995) *Agric. Syst.*, **49**, 337–351.
10. R.A. Fabricant (1970) *J. Regional Sci.*, **10**, 13–24.
11. D. Farrier (1995) *Harvard Environ. Law Rev.*, **19**, 304–305.
12. D.J. Forester and G.E. Machlis (1996) *Conserv. Biol.*, **10**, 1253–1263.
13. J. Gersh (1996) *Amicus J.*, **18**, 14–23.
14. R.P. Greene and J.M. Harlin (1995) *Soc. Sci. J.*, **32**, 137–155.
15. M. Haggerty (1996) Costs of County and Education Services in Gallatin County, Montana. (Bozeman: The Local Government Center), 9 pp.
16. R. Heimlich and M. Vesterby (1992) *Rural Devel. Perspect.*, **8**, 2–7.
17. Yueh-Huang Huang and W.P. Stewart (1996) *J. Travel Res.*, **34**, 26.
18. P.C. Jobes (1988) *Int. J. Environ. Stud.*, **31**, 279–290.
19. H.E. Johansen and G.V. Fuguitt (1990) *Rural Devel. Perspect.*, **2**, 2–6.
20. J.D. Johnson and B.M. Maxwell (1996) *Montana Policy Rev.*, **6**, 22–31.
21. J.D. Johnson and R. Rasker (1995) *J. Rural Stud.*, **11**, 405–416.
22. T.W. Kelsey (1996) *J. Commun. Develop. Soc.*, **27**, 78–89.
23. D. Kemmis (1990) *Community and the Politics of Place*, pp. 44–63. (Norman: University of Oklahoma Press).
24. I.S. Lowry (1966) *Migration and Metropolitan Growth: Two Analytical Models*. (San Francisco: Chandler).
25. P. Nelson and W. Beyers (1998) *Growth & Change*, **29**, 295–318.
26. D.R. Oglethorpe (1995) *J. Agric. Econ.*, **46**, 227–232.
27. H.S. Perloff, E.S. Dunn, E.E. Lampard and R.F. Muth (1960) *Regions, Resources and Economic Growth*. (Galtimore, Maryland: Johns Hopkins Press).
28. D. Pimental, D.A. Takacs, H.W. Brubaker, A.R. Dumas, J.J. Meaney, J.A.S. O'Neill, D.E. Onsi and D.B. Corzilius (1992) *BioScience*, **42**, 354–362.
29. T.M. Power (1995) Thinking About Natural Resource-Dependent Economies: Moving Beyond the Folk Economics of the Rear-View, *A New Century for Natural Resource Management*, edited by R.L. Knight and S.F. Bates, pp. 235–253. (Washington DC: Island Press).
30. M. Rafool and L. Loyacon (1997), *Employment in the Travel and Tourism Industry*, p. 7. (Denver: National Conference of State Legislatures).
31. W.E. Riebsame, W. Gosnell and D.M. Theobald (1996) *Mountain Res. Devel.*, **16**, 395–405.
32. G. Rudzitis, J. Hintz and C. Watrous (1996) Snapshots Of A Changing Northwest. University Of Idaho Department Of Geography, Working Paper From The Migration, Regional Development And Changing American West Project.
33. J.C. Shepard (1993) *Econ. Devel. Quart.*, **7**, 403–410 (1993).
34. D.J. Snepenger, S. Reiman, J.D. Johnson and M. Snepenger (1998) *J. Travel Res.*, **36**, 5–12.
35. T. Theobald (1998) Fragmentation by Inholdings and Exurban Development [Web Page]. Available at: http://www.nrel.colostate.edu:8080//davet/fragmentation.htm.
36. M.G. Turner, D.N. Wear and R.O. Flamm (1996) *Ecol. Appl.*, **6**, 1150–1172.

37. U. S. Bureau of the Census (1995) County Business Patterns, CD-Rom, Web Address http://govinfo.kerr.orst.edu/.
38. D.N. Wear and R.O. Flamm (1993) *Nat. Resour. Model.*, **7**, 379–397.
39. D.W. White, P.G. Minotti, M.J. Barczak, J.C. Sifneos, K.E. Freemark, M.V. Santelmann, C.F. Steinitz, A.R. Kiester and E.M. Preston (1997) *Conserv. Biol.*, **11**, 349–360.

14. MODELLING LAND USE CHANGE WITH LINKED CELLULAR AUTOMATA AND SOCIO-ECONOMIC MODELS: A TOOL FOR EXPLORING THE IMPACT OF CLIMATE CHANGE ON THE ISLAND OF ST LUCIA

Roger White, Guy Engelen and Inge Uljee

INTRODUCTION

Geographical systems are in many respects the most complex phenomena that we confront, because they constitute the nexus of physical, ecological, and human systems. To be usefully understood, they must be treated as an integrated whole–but also in their parts, and in detail, since many of the most important interactions among their components take place locally. The problem of understanding the possible impacts of climate change immediately poses the question of how to deal with this complexity, because impacts are experienced at all scales from global to local, and the causal chains through which they propagate are both multifarious and characterized by numerous feedback loops.

In this chapter we describe a modelling approach that is designed to capture the essentials of the problem by integrating, in a fully dynamic way, models of natural and human systems operating at several spatial scales. The model was developed for the United Nations Environment Program Caribbean Regional Cooperating Unit (UNEP CAR/RCU) (Engelen *et al.*, 1998) with the aim of providing a tool which officials in the region can use to explore possible environmental, social, and economic consequences of hypothesized climate changes. The model is thus generic in the sense that it can accommodate a variety of hypotheses regarding climate change, international economic conditions, and demographic trends, and it can be applied to any small or moderately sized region. The island of St. Lucia was chosen by UNEP as the test application.

INTEGRATED MULTI-SCALE MODELLING

Possible climate change in the Caribbean may be expected to have an important impact on the physical, environmental, social and economic systems of the region (Maul, 1993). Socio-economic systems in the region are already stressed, and climate change may be expected to exacerbate the situation. In St. Lucia, a volcanic island with a rugged landscape, activities are concentrated in the coastal zone. This results in competition for space and conflicts of interest, and causes stress on both terrestrial and marine ecosystems. If changes in climate should occur, or if other non-local events, such as recent changes in the world trade regime, should affect the island, the effects on St. Lucia will not be evenly felt. Some economic or demographic sectors will be affected much more dramatically than

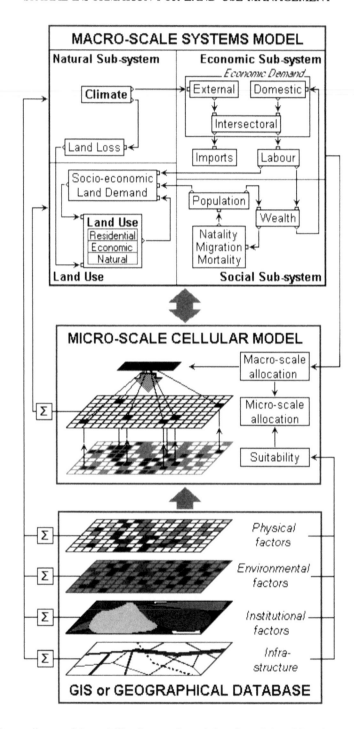

Figure 1. System diagram of the modelling framework consisting of coupled models at the macro- and micro- and GIS levels.

others and some parts of the island may experience a positive impact, while others will be negatively affected, or perhaps not affected at all. Climate change in the Northern Hemisphere, for example, could reduce demand for tourism on the island, with effects that would be felt throughout the economy; economic difficulties would in turn affect population growth through, for example, changes in migration. These economic and demographic changes would then have impacts on land use (Plate 23 top). If worsening economic conditions were not offset by increased out-migration, then in some areas forests or other ecologically valuable land covers could be converted to agriculture, with, in turn, an additional negative impact on tourism.

To capture the response of a complex and integrated system such as this to a forcing factor like climate change requires that all important components be modelled, as well as the links between the components. Hence the model described here integrates several sub-models operating at two spatial scales (Figure 1).

At the macro-scale, essentially the scale of the entire island, several components are specified. First, a module for developing climate change scenarios is augmented with a tool for specifying hypotheses concerning relationships between changes in climate parameters and changes in certain economic variables such as demand for tourism and agricultural output. This is linked through the economic variables to an intersectoral (Input-Output) model of the economy, which is in turn linked to a demographic model. These three models, both separately and jointly, implicitly entail changes in land use — the climate module through changes in sea level, the economic model through changes in output (which require land to be realised), and the demographic model through changes in land required for housing. But they say nothing about where the land use changes will occur, or indeed whether the changes are even possible given constraints on the amount and quality of land available.

It is at the micro-scale that the locational aspects are modelled. This is done by means of a cellular automata based model of land use dynamics which makes use not only of land use data, but also of other relevant spatial data such as soil type, slope, precipitation, and transportation infrastructure. This model is driven by the demands for land that are generated by the macro-scale model, so it is able to show the spatial manifestation of the economic and demographic impacts of climate change. Furthermore, some micro-scale spatial data is returned to the economic and demographic models where it modifies parameter values, thus enabling those macro-models to reflect the particular local conditions that affect productive activities.

This general approach, in which diverse models of human and natural systems are linked through a cellular land use model, has been implemented not just in St. Lucia, but in several other problem contexts as well. One, for example, focuses on the interrelated dynamics of metropolitan expansion, agricultural land use change, and regional hydrology in Southwest Sulawesi, Indonesia (Uljee et al., 1999). Another models the spatial dynamics of the Dutch population and economy down to the level of land use changes and their attendant ecological impacts; it was developed as a tool for use in the development of national land use policies (White and Engelen, 1999).

The integration of models covering several problem domains is essential to this approach. Yet other than the applications just mentioned, the area of integrated modelling offers surprisingly few examples in which models of natural and human systems are connected dynamically. Among the few are Bockstoel et al. (1995) and Rotmans et al.

(1994). More common are approaches in which models are chained, without the possibility of dynamic feedback, for example models in which climate change drives changes in agriculture, but the latter have no effect on climate (e.g. Parry and Carter, 1989; Strzepek and Smith, 1995). One reason that integrated modelling is so uncommon may be that many of the relevant domain models are non-spatial, while human and natural systems tend to interact in a place specific context. Thus integration is facilitated when models are made spatial, or given a spatial interface. In this respect cellular automata constitute the heart of integrated modelling: they are dynamic, spatial, and multivariate, and thus provide an ideal interface for linking models of different phenomena.

CELLULAR AUTOMATA BASED MODELLING OF LAND USE DYNAMICS

Introduction to Cellular Automata

Tobler (1979) first proposed using cellular automata (CA) as a tool for modelling spatial dynamics. Couclelis (1985, 1988, 1997) explored the implications of the idea in an important series of theoretical papers. The approach has since been implemented by others in variety of applications (e.g. Batty and Xie, 1994; Benenson, 1998; Ceccini and Viola, 1990; Clark et al., 1997; Papini and Rabino, 1997; Phipps, 1989; Portugali and Benenson, 1995; White and Engelen, 1993, 1999; White et al., 1997; Xie, 1996). And the approach has been linked to GIS (Wagner, 1997; White and Engelen, 1994; Wu, 1998).

Cellular automata can be thought of as very simple dynamic spatial systems in which the state of each cell in an array depends on the previous state of the cells within a neighbourhood of the cell, according to a set of state transition rules. Because the system is discrete and iterative, and involves interactions only within local regions rather than between all pairs of cells, a CA is very efficient computationally. It is thus possible to work with grids containing hundreds of thousands of cells. The very fine spatial resolution that can be attained is an important advantage when modelling land use dynamics, especially for planning and policy applications, since spatial detail represents the actual local features that people experience, and that planners must deal with.

A conventional cellular automaton consists of

1) a *Euclidean space* divided up into an array of identical cells;
2) a cell *neighbourhood* of a defined size and shape;
3) a set of discrete *cell states;*
4) a set of *transition rules,* which determine the state of a cell as a function of the states 5) of cells in the neighbourhood;
6) and *discrete time steps*, with all cell states updated simultaneously.

However, these defining characteristics can be interpreted broadly, or relaxed in response to the requirements of a particular modelling problem, so many types of CA are possible.

Cellular automata offer a number of advantages. As already mentioned, they are computationally efficient and so, unlike traditional regional models, permit extreme spatial detail. They are thus able to reproduce the actual complexity, frequently fractal in nature, of real spatial distributions. Furthermore, because of the high resolution and raster nature, they are compatible with GIS databases, and can be linked with them in a relatively straightforward way. At the other end of the spatial scale, CA can be linked through their transition rules to other, macro-scale models that constrain or drive the CA dynamics. This facilitates comprehensive modelling of integrated environmental-human systems. Finally, CA are defined and calibrated in a single operation, since calibration amounts to finding the optimal transition rules. This means that CA are typically implemented much more quickly and easily than traditional spatial models. These advantages are illustrated in the CA model of land use dynamics in St. Lucia.

A Cellular Automata Model for St Lucia

In the CA land use model of St. Lucia, land uses and land covers are represented by cell states, so the CA dynamics models the year-by-year changes in land use and land cover. For each cell on the map, the cell neighbourhood is examined, and on the basis of the land uses and covers within the neighbourhood, a set of values, or *potentials,* is calculated, one value for each possible land use that could occupy the cell. These values represent the desirability of the cell for each of the possible land uses, given the composition of the neighbourhood. They thus capture local attraction and repulsion effects. In addition, the inherent suitability of the cell itself for each activity is determined; these suitabilities represent such characteristics as elevation, slope of the land, and soil type. Also, the accessibility of the cell to the transport network is calculated, and weighted according to each of the potential land uses, since for some activities accessibility is more important than for others. Finally, considering these three factors, the cell is changed to the land use for which is most desirable — but only if there is a demand for that land use, and these demands are generated outside the CA. The St. Lucia CA is specified as follows:

The cell space

The island of St. Lucia and the surrounding coastal waters are covered by 22 506 cells arranged in a matrix of 186 rows by 121 columns. The resolution (cell size) is 250 m., which accords well with available land use/land cover data, but does not permit a very good representation of the terrain.

Unlike traditional CA, the cell space is not assumed to be homogeneous. In order to represent the real variations in land quality, legal and administrative constraints on land use, and variations in the accessibility of land, the CA is run on a cell space in which each cell has an intrinsic suitability for each possible land use (Plate 23 bottom). The suitability of a cell for a particular land use is a measure of the capacity of the cell to support that activity. In the St Lucia model, suitabilities are calculated as a linear combination of a series of physical, environmental, and institutional factors such as topography, soil quality, rainfall, sensitivity to erosion, planning regulations, and other legal restrictions on land use; they are normalized so that all values fall within a range of zero (completely unsuitable) to one. Suitability values are treated as constants, and thus are not altered dynamically during execution of the land use model; values may change, however, as a result of user intervention or sea level changes, as explained below. The suitability calculations are performed outside the model, in a GIS — *Idrisi* in this case.

Another cell space inhomogeneity is related to the accessibility of a cell to the rest of the region. The accessibility value of a cell for a particular function is a measure of the importance of proximity to the road network for that function (Plate 23 bottom). The accessibility factor of a cell for land use z is calculated as:

$$A_z = \frac{1}{\left(1 + \dfrac{D}{a_z}\right)} \qquad (1)$$

where D is the Euclidean distance from the cell to the nearest cell on the road network and a_z is a parameter expressing the importance of access to a road for activity z; it is calibrated separately for each activity. The equation returns accessibility values that are in the range (0–1).

The neighbourhood

The cell neighbourhood is a circular region eight cells in radius, containing 196 cells falling into 30 discrete distance bands (1, $\sqrt{2}$, 2, $\sqrt{5}$,...). This relatively large neighbourhood, with a radius of two kilometers, probably comes close to encompassing the locally perceived neighbourhood to which locational agents respond.

The cell states

Fifteen land use and land cover categories are used, each represented by a cell state. These are divided into two categories: *functions*, or active states, for which CA transition rules are defined and which thus have a dynamic, and *features*, which are cell states are fixed. While features states cannot be changed by the transition rules, they may appear as arguments in those rules, and thus affect the dynamics of the function states. Furthermore, they may be changed by other rules outside the CA proper, or by user intervention.

The land use/land cover states are shown in the following table:

Functions:	Features:
natural vegetation (mainly bush)	forest reserve
forest (mainly secondary forest)	mangroves
agriculture	sea
industry and quarries	beach
trade and services	coral reef
tourism	terminals, ports, airports, *etc.*
rural residential	infrastructure, water, electricity
urban residential	

The transition rules

Unlike traditional cellular automata, the transition rules are grouped into three classes which establish the priority with which they are executed. This is necessary because the

CA proper cannot handle all the types of state transitions which must be included in the St. Lucia model, reflecting the fact that the model is designed to provide functionality required by the end users, rather than simply to explore the use of CA.

Rules of priority one represent user interventions. They are used to introduce hypothetical planning decisions during the execution of the model, so that the user may run what-if experiments. They include interventions such as imposing a land use on a specific cell (*e.g.* an airport extension); converting a water-covered area to land (*e.g.* a land fill project) by editing elevations in the digital elevation map; or adding a transportation link to change the relative accessibility of certain areas. These rules apply only to user designated cells, and, in the case of land use changes, apply *as rules* only for the time step in which they are made; after that, the system either accepts or rejects them on the basis of its endogenous dynamics. A rule of priority one overrides all other possible changes made by rules of priority two or three.

Rules of priority two implement land use/land cover changes caused directly by changes in sea level. With a rising sea level, cells that fall below certain threshold elevations must be converted, for example, from forest to mangrove, or from beach to sea. With a falling sea level, emerging land must be assigned a land cover; this is done on the basis of user-defined suitabilities. The following set exemplifies these rules:

$$if \ (h < 0) \ then \ Z_{sea}$$

$$if (0 \leq h < 0.5) \ then \left[if \left(\sum_{N8} Z_{mangrove} > \sum_{N8} Z_{beach} \right) then \ Z_{mangrove} \ else \ Z_{beach} \right] \quad (2)$$

$$if \ (h \geq 0.5) \ then \ [if \ (S_{forest} > S_{natural}) \ then \ Z_{forest} \ else \ Z_{natural}]$$

where h is the elevation, Z is the land use/land cover, $N8$ is the eight-cell radius neighbourhood, and S is the suitability for the indicated land use.

Rules of priority three apply to the Function states listed above. They constitute the heart of the CA land use model. Except for "Natural" and "Forest", which are residual or "no-use" states, these are the land uses that correspond to state variables in the Macroscale model. For rules of priority three it is first necessary to calculate for each cell a vector of transition potentials, one for each activity. The transition potential is an expression of the 'strength' with which a cell is likely to change state to the state for which the transition potential is calculated. Transition potentials are calculated as weighted sums:

$$P_z = f(S_z).f(A_z). \sum_d \sum_i (w_{z,y,d} \times I_{d,i}) + \varepsilon_z \quad (3)$$

where P_z = potential for transition to state z,

$f(S_z)$ = function $(0 \leq f(S_z) \leq 1)$ expressing the suitability of the cell for activity z,

$f(A_z)$ = function $(0 \leq f(A_z) \leq 1)$ expressing the relative accessibility of the cell for activity z,

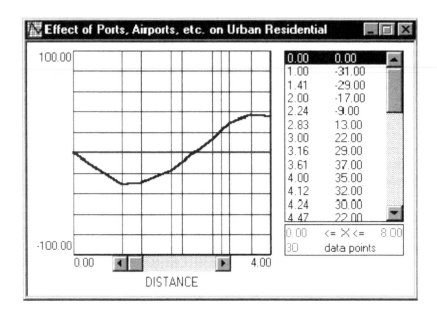

Figure 2. The function editor used to enter the weighting parameters applied to the cell neighborhood; the graph shows the influence of the proximity of ports *and airports* on the desirability of a cell for *urban residential* use.

$$f(S_z) = S_z; \text{ if } \sum_d \sum_i (w_{z,y,d} \times I_{d,i}) > 0 \qquad (4)$$

$$f(S_z) = (1 - S_z); \text{ if } \sum_d \sum_i (w_{z,y,d} \times I_{d,i}) < 0 \qquad (5)$$

where $w_{z,y,d}$ = the weighting parameter applied to cells with state y in distance zone d
        $(0 \leq d \leq 30)$,

   i = the index of cells within a given distance zone d

   $I_{d,i}$ = the Dirac delta function: $I_{d,i} = 1$ if the state of cell i in distance zone d
        is y; otherwise $I_{d,i} = 0$

   ε_z = a stochastic disturbance term, with $\varepsilon_z = 1 + [-\ln(rand)]^a$

Thus, cells within the neighbourhood are weighted differently depending on their state y and also depending on their distance d from the center cell for which the neighbourhood is defined (Figure 2). Since different parameters can be specified for different distance zones, it is possible to build in weighting functions that have distance decay properties similar to those of traditional spatial interaction equations. The deterministic transition potentials are subjected to the stochastic perturbation, ε_z, to account for individual preferences and unknown factors in location decisions.

Values of the weighting parameters $w_{z,y,d}$ represent the total weight accorded to the cells of a particular land use within a distance ring. Negative values express push or repulsion effects, while positive numbers express pull or attraction effects. The weighting coefficients are relative numbers; in other words, what is important is the size of the weights given to various land uses at various distances within the neighbourhood, relative to each other, for calculation of the potential for transition to a particular land use, and also the sizes of these weights relative to those used in the calculation of transition potentials for other land uses.

The transition potentials calculated for each cell, one for each functional land use, represent the propensies of the cell to convert to each of these states (Plate 23 bottom). The transition rule is then to convert each cell to the state for which it has the highest potential — subject, however, to the constraint that the number of cells in each state be equal to the demand for cells established by the Macro-scale model. More specifically, to select the T_z cells to receive the function z at each iteration, the potentials calculated for all cells for transition to all states are ranked from highest to lowest. Starting with the highest value from this list, the T_z cells with highest potentials for transition to each state z are identified and the transitions are executed. Hence, a cell will change to the state for which its potential is highest, unless it is not among the T_z highest potentials, in which case it might change to a state z' for which its potential ranks second, and so forth. All the cells for which potentials are not among the T_z highest for any of the z states will remain in or return to either the *natural* or *forest* state, depending on the suitabilities.

DRIVING THE CELLULAR AUTOMATON: LINKED MACRO-SCALE MODELS

The Natural Sub-System: Specifying Hypotheses

The natural sub-system of the macro-scale model ensemble consists of a set of linked relations expressing the change through time of temperature and sea level, and the effects of these changes on precipitation, storm frequency and, secondarily, external demands for services and products from St. Lucia. This component is not a true model, but rather a set of linked hypotheses which the user is free to change.

In particular, it is assumed that exogenous changes in mean annual temperature affect precipitation and storm frequency, and that these three variables in turn drive changes in sectoral output of the economy:

$$P_t = f_P(T_t) \quad \text{and} \quad F_t = f_F(T_t), \quad \text{with} \quad T_t = f_T(t) \tag{6}$$

$$E_{i,t} = e_{i,t} [1 + f(T_t) + f(P_t) + f(F_t)] \tag{7}$$

where t = time
T_i = temperature change
P_i = precipitation change
F_i = storm frequency change
$E_{i,t}$ = vector of sectoral external demands;
$e_{i,t}$ = vector of projected sectoral external demand, *ceteris paribus*

In addition, the exogenous hypothesized change in sea level causes changes in beach area, and this, together with the climate variables, is assumed to affect the demand for tourism, since beach area in St. Lucia is limited while tourism is heavily beach oriented:

$$E_{tourism,t} = e_{tourism,t} \times \frac{1 - \beta \times \exp(-\alpha \times N_{beach,t})}{1 - \beta \times \exp(-\alpha \times N_{beach,t=0})} \times [1 + f(T_t) + f(P_t) + f(F_t)] \qquad (8)$$

where $L_t = f_L(t)$ = sea level change
        $N_{beach,t}$ = number of beach cells
        α = resilience of tourism industry to changing beach area
        β = relative importance of beach tourism in total tourism industry

The relationships described by equations. 6–8 are obviously strong simplifications of the actual phenomena. In principle, they could be replaced by more elaborate climate models, but realistic climate models are notoriously complex; and while models of the response of agricultural productivity to climate changes are available, for other economic sectors no such models exist. Since the purpose of the St. Lucia model is to provide a tool for exploring the impacts of climate change, rather than to model climate change itself, it seems reasonable to treat these forcing relationships as user-modifiable hypotheses. Nevertheless, the specific default relationships used in the model reflect discussions with members of the UNEP/IOC Task Team on the Implications of Climatic Change in the Wider Caribbean Region.

The Demographic Model

The demographic model calculates the population in St. Lucia at yearly intervals on the basis of estimated births, deaths, and net migration. It is extremely simple in that it is not disaggregated by sex or age cohort; but on the other hand, it does include features that allow it to capture the effects changing economic conditions. Birth, death, and net migration rates are each assumed to be characterized by a secular structural trend; these trends can be defined by the user. In addition, in the case of the mortality and migration rates, the structural component is supplemented by an economic component. While the structural component represents the long term, underlying change in the rate, the economic component captures changes in the structural rate due to changes in economic trends. The economic component is particularly important in the case of migration, since the volume of migration responds quickly to changes in relative economic conditions. Initial values of birth, death, and migration rates are taken from statistics published by the Government of St. Lucia; structural trends in the mortality and migration rates are estimated from time series data in the same sources (Government of St. Lucia, 1992a, 1992b, 1993).

$$bb_t = f_b(t) \qquad (9)$$

$$dd_t = 0.5r_2 f_d(t)[1 + \exp(-r_3(u_t - u_{t=0}))] \qquad (10)$$

$$mm_t = r_4 + \frac{[1 - r_5 \exp(u_t - u_{t=0})]}{[1 - r_5]} - 1 \qquad (11)$$

$$\Delta p_t = p_t(bb_t - dd_t - mm_t) \tag{12}$$

$$p_{t+1} = p_t + \Delta p_t \tag{13}$$

where p_t = population
bb_t = birth rate
dd_t = mortality rate
mm_t = migration rate
u_t = employment participation index, calculated as the sum of sectoral production data (generated by the economic model), divided by the population.
$f_b(t)$ = structural reproduction rate
$f_d(t)$ = structural mortality rate
r_2 = structural death rate
r_3 = variable mortality rate
r_4 = structural migration rate
r_5 = variable migration rate

The Economic Model: An Input-Output Approach

The economy of St. Lucia is modelled by means of a highly aggregated input-output (I-O) model (Figure 3) which is coupled to the demographic sub-model which describes the economy of the island as a set of linear equations. This approach has the advantage of ensuring that the output of the economic model is internally consistent. In addition, it captures the interdependencies between economic sectors and thus well represents the multiplier effects by which changes in one sector propagate through the entire economy. On the other hand, the method is based on an assumption of constant technical coefficients, so that changes in the underlying structure of the economy, due, for example, to technological innovations, import substitution, or factor substitutions in response to changes in relative prices are not represented. This is not a serious problem for short run situations, but becomes much more of a concern when the modelling period runs to several decades. Ideally, in the context of the present model, the technical coefficients would evolve in response to changes in productivity generated by the cellular model. For example, if agricultural activity is forced onto less suitable land, the output per unit of labour, as represented by a technical coefficient in the I-O model, should decline.

The I-O model is used in a quasi-dynamic manner, since at each iteration the final demand sectors change exogenously: the domestic demand changes in response to population growth or decline, and exports change in response to both secular trends and the climate changes as described in equations. 7 and 8. The equations must therefore be solved at each iteration. In turn, the output of the I-O model affects the demography (equations. 10 and 11), and through changes in the amount of land required to accommodate changes in sectoral production, it affects also the land use dynamics of the cellular model. The actual I-O table for 1990 was estimated from economic data published by the Government of St. Lucia (Government of St. Lucia, 1992a, 1992b, 1993). It consists of five industry sectors and two final demand sectors. The industry sectors — agriculture, industry, trade, services, and tourism — correspond, after combining trade and services into a single category, to the four economic functions in the cellular model. The final demand sectors

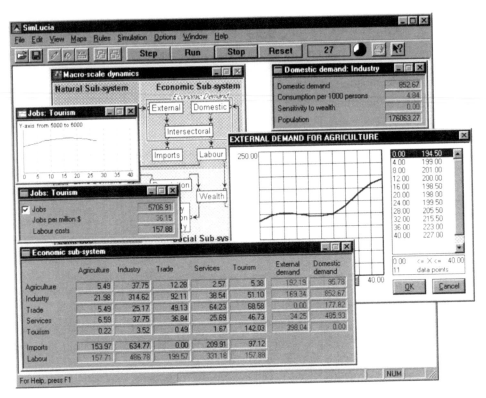

Figure 3. Graphical user interface for the economic sub-system, showing the input-output table and several other components.

— domestic final demand and exports — are exogenous, and drive the I-O model. Domestic final demand is a function of population and the employment participation rate, while exports are given by a user defined temporal trend. The model is as follows:

$$\Delta Y_{i,t} = \Delta Dom_{i,t} + \Delta E_{i,t} \tag{14}$$

$$Y_{i,t} = Y_{i,t-1} + \Delta Y_{i,t} \tag{15}$$

$$\Delta S_{i,t} = \sum_j A_{i,j} \Delta S_{j,t-1} + \Delta Y_{i,t} \tag{16}$$

$$\Delta S_{i,t} = \Delta S_{i,t-1} + \Delta S_{i,t} \tag{17}$$

$$\Delta M_{i,t} = I_i \times \Delta S_{i,t} \tag{18}$$

$$M_{i,t} = M_{i,t-1} + \Delta M_{i,t} \tag{19}$$

$$\Delta X_{i,t} = \frac{\Delta S_{i,t}}{B_i} \tag{20}$$

$$X_{i,t} = X_{i,t-1} + \Delta X_{i,t} \tag{21}$$

where $S_{i,t}$ = vector of total sectoral outputs
$\quad\quad\;\; Y_{i,t}$ = vector of final demands
$\quad\quad\;\; X_{i,t}$ = vector of sectoral employment
$\quad\quad\;\; M_{i,t}$ = vector of sectoral imports
$\quad\quad\;\; Dom_{i,t}$ = vector of sectoral domestic demands
$\quad\quad\;\; E_{i,t}$ $= f_e(t)$ = vector of sectoral external demands
$\quad\quad\;\; A_{i,j}$ = matrix of technical coefficients for endogenous sectors
$\quad\quad\;\; I_i$ = vector of import coefficients
$\quad\quad\;\; B_i$ = vector of employment coefficients

The employment coefficients B_i representing output per employee, were calculated from 1990 sectoral output and employment data. Further,

$$E_{i,t} = f_e(t) \tag{22}$$

and

$$\Delta Y_{i,t} = \Delta p_t c_i \exp(n_i(u_t - u_{t=0})) + \Delta E_{i,t} \tag{23}$$

where u_t = employment participation index
$\quad\quad\;\; p_t$ = total population
$\quad\quad\;\; c_i$ = vector of sectoral domestic demand coefficients
$\quad\quad\;\; n_i$ = vector of influences of employment index on domestic consumption

Equation 23 expresses changes in domestic demand for the economic sectors in terms of the change in population and the consumption *per capita* of the sectoral product. Consumption *per capita* in turn depends on the relative employment level, reflecting the fact that people spend differing proportions of their budget on each type of good as their economic status changes.

Land Productivity Calculations: The Link to the Cellular Model

Changes in economic activity and population entail changes in the amount of land required to carry out the activity or to house the people. Thus changes in sectoral output and population must be translated into demands for land, using current land productivity or density levels. The new demands for land are then passed to the cellular model, which then allocates the required land (Plate 24).

Density is defined as the number of people who can live or work on one cell. It varies through time primarily as a function of the demand for land relative to its availability — i.e. the scarcity of land. Scarcity is defined in terms of the amount of land occupied by an activity as a proportion of the total amount of land available for the activity, with the latter being defined as the land already occupied by the activity together with land in the *natural* and *forest* states; both quantities are weighted by the suitability of the land. The greater the scarcity of suitable land, the higher the density, *ceteris paribus*. Density also depends directly on the mean suitability of the land occupied by the activity: the more suitable the land, the higher the density. Specifically,

$$W_{i,0} = \frac{X_{i,0}}{N_{i,0}} \tag{24}$$

$$W_{i,t} = W_{i,0} \left(\frac{SS_{i,t}}{TS_{i,t}} \frac{TS_{i,0}}{SS_{i,0}} \right)^{\sigma_i} \left(\frac{SS_{i,t}}{N_{i,t}} \frac{N_{i,0}}{SS_{i,0}} \right)^{\zeta_i} \tag{25}$$

where the subscript i includes both residential and economic activities, and

> $W_{i,t}$ = density of activity i
> $SS_{i,t}$ = suitabilities for activity i summed over the cells occupied by the activity
> $TS_{i,t}$ = suitabilities for i summed over all natural and forest cells
> $N_{i,t}$ = total cells required for activity i
> $X_{i,0}$ = initial sectoral employment
> σ_i = sectoral sensitivities to land pressure
> ζ_i = sectoral sensitivities to land quality

Finally, the total land that is required for each activity is calculated, and these demands are then passed to the cellular model; thus the loop between macro- and micro-models is closed:

$$N_{i,t+1} = \frac{X_{i,t}}{W_{i,t}} \tag{26}$$

$$N_{p,t+1} = \frac{p_t}{W_{i,t}} \tag{27}$$

where p_t = total population and other symbols are as above.

In addition, while the model is running, the suitability values for all cells actually occupied by a particular land use are monitored. Changes in the total and average suitability for each land-use are passed back to the macro-level and result in further changes in the land density variables. In other words, the detailed land suitabilities and land-use patterns in the micro-level model have their effect in processes that are modelled at the Macro-level. The geography at the micro-level thus affects the global dynamics directly and continually.

DISCUSSION AND CONCLUSIONS

It might seem that linking a number of models, each with its own uncertainties and inadequacies, would result in a multiplication of errors to the point where the output of the integrated model would be useless. That does not seem to be the case. Since the models are generally complementary, each tends to limit the size of errors that the others can produce. For example, a runaway population model might try to put an impossible number of people in the region. But before that could happen, feedback from the economic model would increase the out-migration rate to limit the population. The economic model would in turn be responding to information on land availability and suitability fed to it from the

CA and the GIS. Experiments with a model which operates at three spatial scales (national, regional, and cellular) show that errors in both the macro-scale economic-demographic model and the CA land use model are reduced substantially (on the order of 70% for the former) when they are linked (White and Engelen, 1999).

The integrated modelling approach described in this chapter brings together not only different models but also different modelling techniques. The macro-scale models represent the system dynamics tradition of modelling using differential or difference equations. At the micro-level, GIS capabilities are used to model the suitabilities and other spatial inhomogeneities that characterize the cell space — which is essentially the same as a GIS raster space. The CA itself might thus be considered, in a sense, to be a fully dynamic GIS. This "dynamic GIS" interpretation of the CA makes explicit the fact that the spatial patterns stored and manipulated in a GIS are typically dated: in general they represent an unstable situation that will evolve toward an equilibrium configuration. But that final state is not reached, because the macro-scale models continuously displace the equilibrium that the CA is trying to find. Similarly, the dynamics of the CA displace the equilibrium which the macro-scale models are trying to reach. This propagation of disequilibrium exemplifies the non-linearity of the linked models.

The non-linearity (together with the stochasticity) allows the model to capture explicitly two aspects of the inherent unpredictability of the world. It also means that the system trajectories are subject to bifurcations. In other words, near critical points, small changes in parameter values or initial conditions, as well as the stochastic fluctuations, can alter the future of the system substantially. Bifurcations are the key to proper use of this type of model. They define domains in parameter space where the results are qualitatively similar regardless of the precise values of the parameters. Once the limits of the domain are passed, however, the behaviour of the model, and thus the results generated, will change noticeably. So while we can never know precisely what the future will bring, the model will give a usefully reliable indication that if current conditions are like *this*, then the situation in ten years will probably be substantially like *that*. Such general indications of future conditions are potentially quite useful to planners and policy makers.

SOFTWARE

A fully functional demo version of the SIMLUCIA Software as well as a manual, can be downloaded from http://www.riks.nl.

ACKNOWLEDGEMENTS

We gratefully acknowledge the assistance of Anthony Philpott and Dr. Alvin Simms of the Geoidal Laboratory of the Department of Geography, Memorial University of Newfoundland, who digitized and further processed the numerous maps of St. Lucia.

REFERENCES

1. M. Batty and Y. Xie (1994) *Environ. Plan. B*, **21**, 31–48.
2. Benenson (1998) *Comput., Environ. Urban Syst.*, **22**, 25–42.
3. N. Bockstoel, R. Costanza, I. Strand, W. Boynton, K. Bell and L. Wainger (1995) *Ecol. Econ.*, **14**, 143–159.
4. Cecchini and F. Viola (1990) *Wissenschaftliche Zeitschrift der Hochschule fur Architektur und Bauwesen*, **36**, 159–162.
5. K. Clark, S. Hoppen and I. Gaydos (1997) *Environ. Plan. B*, **24**, 247–261.
6. H. Couclelis (1985) *Environ.Plan. A*, **17**, 585–596.
7. H. Couclelis (1988) *Environ. Plan. A*, **20**, 99–109.
8. H. Couclelis (1997) *Environ. Plan. B*, **24**, 165–174.
9. G. Engelen, I. Uljee and R. White (1998) *Vulnerability Assessment of Low-lying Coastal Areas and Small Islands to Climate Change and Sea Level Rise.* (Maastricht: Research Institute for Knowledge Systems).
10. Government of St. Lucia, Govt. Statistics Dept. (1992a) *Annual Statistical Digest 1991.* (Castries).
11. Government of St. Lucia, Govt. Statistics Dept. (1993) *National Accounts 1977 to 1992.* (Castries).
12. Government of St. Lucia (1992b) Min. of Agriculture, Lands, Fisheries and Forestry, *Annual Agricultural Statistical Digest 1992.* (Castries).
13. G. Maul (editor) (1993) *Small Islands. Marine Science and Sustainable Development, Coastal and Estuarine Studies*, **51**. (Washington DC: American Geophysical Union).
14. L. Papini and G. Rabino (1997) *ACRI '96: Proceedings of the Second Conference on Cellular Automata for Research and Industry*, pp. 147–157. (Berlin: Springer).
15. M. Parry and T. Carter (1989) *Clim. Change*, **15**, 95–116.
16. M. Phipps (1989) *Geog. Anal.*, **21**, 197–215.
17. J. Portugali and I. Benenson (1995) *Environ. Plan. A*, **27**, 1647–1665.
18. J. Rotmans, M. Hulme and T. Downing (1994) *Global Environ. Change.*
19. K. Strzepek and J. Smith (1995) *As Climate Changes: International Impacts and Implications: An Assessment of Integrated Climate Change Impacts on Egypt.* (Cambridge University Press).
20. W. Tobler (1979) *Philosophy of Geography*, pp. 379–386. (Dordrecht: Reidel).
21. Uljee, G. Engelen and R. White (1999) *Integral Assessment Module for Coastal Zone Management: RamCo 2.0 User Guide.* (Maastricht: Research Institute for Knowledge Systems).
22. D. Wagner (1997) *Environ. Plan. B*, **24**, 219–234.
23. R. White and G. Engelen (1993) *Environ. Plan. A*, **25**, 1175–1199.
24. R. White and G. Engelen (1994) *Geog. Syst.*, **1**, 237–253.
25. R. White and G. Engelen (1999) *Comput. Environ. Urban Sys.*, in press,.
26. R. White, G. Engelen and I. Uljee (1997) *Environ. Plan. B*, **24**, 323–343.
27. F. Wu (1998) SimLand: a prototype to simulate land conversion through the integrated GIS and CA with AHP-derived transition rule. *Int. J. Geog. Inf. Sci.*, **12**, 63–82.
28. Y. Xie (1996) *Geog. Anal.*, **28**, 350–373.

15. CONCLUSION: A FRAMEWORK FOR USE OF SPATIAL INFORMATION IN ANALYSIS AND MODELLING FOR LAND USE MANAGEMENT AND PLANNING

Richard J. Aspinall

INTRODUCTION

This chapter provides a synthesis of the examples presented in the chapters of the book. The aim is to present a framework for developing and applying data, analyses and modelling in order to be inclusive of the multiple perspectives and knowledge needed for integrated land use management and planning. The integration of these scientific tools with decision-making processes are also discussed. The issues identified in the introductory chapter are revisited as part of this synthesis.

Frameworks for applying spatial information to land use management and planning are typically discussed in context of Decision Support Systems (DSS); this approach is followed here. Decision support systems are usually considered a set of technologies that include data, models, and GIS, the spatial component from the GIS identifying Spatial Decision Support Systems. The activities of predicting land use change and developing information to support decisions in land use management and planning are only partially technical. Land use management and planning are creative design activities that take place within prevailing social, legal, political, and economic contexts as well as being concerned with particular resources within specific geographic areas. The role of decision support technology and spatial information science developed for application to land use lies, therefore, within this social process of creation and design and the technological aspects of decision support are most properly designed and developed within this social context. A goal of a spatial decision support system in place-based decision support is thus to facilitate collaboration and communication as part of a process and thus to play an important social role beyond their construction as technology.

Many decisions in land use management and planning are concerned with management of change and design or selection of alternative future directions for land use. The framework presented follows this design process, finding a role for scientific databases, analyses, models and other aspects of spatial data handling and analysis at important stages of the process. This role is identified within a wider role for spatial scientists than the traditional one of objective and detached observer or technologist. In the creative processes of decision-making, land use management and planning, the scientist must become an active participant in a process that is fundamentally social.

Some of the main characteristics of emerging approaches to integrated resource management are described by Bellamy *et al.* (1999) and are summarized in Table 1. These characteristics emphasize political and social context and actions, concern for larger spatial

Table 1.   Characteristics of emerging approaches to integrated resource management
(after Bellamy *et al.*, 1999)

UNDERLYING PRINCIPLES:
Multiple use of resources
Link human and environmental systems
Focus on equity and social justice
Maintain and enhance resource condition and long-term sustainability

APPROACH:
Holistic management involving natural and human systems
Management over a long-term perspective (temporal scale)
Management over large geographic areas (spatial scale)
Best management practice based on economic, environmental and social trade-offs
Interaction and collaboration are adaptive and iterative throughout the decision making process
Interaction and collaboration are focused on facilitated learning throughout the decision making process
Recognition of a broad range of meaning and use for landscapes and resources based in diversity of
cultural history and place-based values among participants
Scientists and technology are participants, not solutions or independent 'experts'
Technology plays a social role in the participative R&D process

METHOD:
Decision-making is dominated by political considerations
The decision process is inclusive, open and involves multiple stakeholders and public participation
Research and development is collaborative and participatory
Decisions emerge from the whole collaborative, multidisciplinary, social, and technical process of research
and development.

scales, longer time scales, multiple use of resources with multiple goals, and decision making through an inclusive process of collaboration and participation involving multiple disciplines, perspectives and stakeholders.

In simplified form, decision-making processes follow six main stages as outlined in Figure 1. There are continually feedbacks between each of these stages and the process is by no means as linear as Figure 1 suggests. Figure 1 also shows some of the principle topics for further development and application of spatial database, management and analysis systems at each stage of the decision process. Developing spatial information as part of this whole process of land use management and planning requires advances in four main areas:

1) A need for improved understanding and framing of the issues involved in land use management. What are the issues involved? What is the contribution of spatial information at different stages of the process?

2) Focusing technological and methodological developments on the data, tool and information needs of land use decision-makers within the decision process. What information is needed to facilitate decision-making? How can technology facilitate wide participation in, and during, the decision-making process?

3) Developing a greater awareness and understanding of decision-making processes, including information needs and use, and the timescales of decision-making. How can development of decision support be part of the decision process? How can develop-

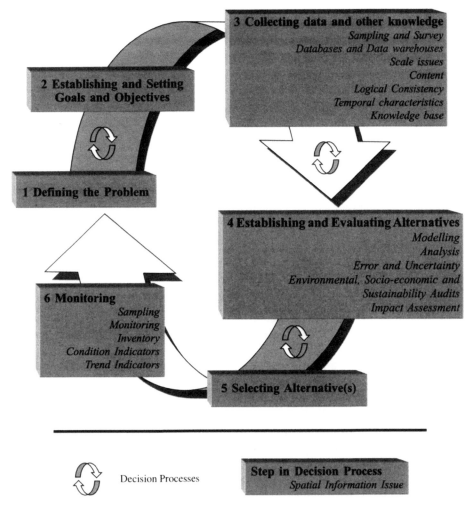

Figure 1. Stages in the decision process. The different stages are linked by feedbacks that make the process complex and seldom followed in strict sequence. Spatial information issues associated with each stage are also identified.

ment be inclusive? How can the research and development be integrated with the decision process?

4) Developing monitoring and feedback systems that more tightly link the consequences of decision-making with the information provision and analytical processes of technical and methodological developments. How do monitoring and evaluation link with the goals and objectives of a decision? How is the comparison and evaluation of alternatives carried out in an objective and inclusive way?

Focusing on these uses of technology as an integral part of the social process of decision-making will facilitate both technical improvements in data handling and information processing and also lead to decision support systems that are collectively the 'property' of the decision makers and communities.

ISSUES IN LAND USE MANAGEMENT

The chapters of this book address a wide range of land use issues and management needs. Although some of the specific examples focus on single land use issues, the goal of each of the chapters is to contribute to the more general objective of balancing the use of resources for multiple goals and outcomes. Chapters 13 (Maxwell *et al.*) and 14 (White *et al.*) specifically address multiple-issue, multiple land use questions and develop models, visualization and other tools and approaches that directly attempt to address the balance of these issues and varied drivers of land use change. Issues faced in land use management and planning increasingly require decision-makers to adopt holistic approaches that attempt to consider large geographic areas and longer time scales for multiple use of resources. This focus is partly based in the concepts of sustainability, social justice, intergenerational equity, and recognition that best management practice requires trade-offs between economic, environmental and social goals; it is also a necessary concern as a consequence of increasing human pressures and demands on natural resources as populations and economies develop and environmental conditions and technology change.

Setting goals and objectives for a land use decision as part of the specification of the problem being addressed occupies the early stages of the decision process (Figure 1: Stages 1 and 2) and provides a foundation for all the subsequent stages outlined in Figure 1. In Chapter 11, Conroy distinguishes goals as broad statements of a desired consequence resulting from accumulated decisions, and objectives as mathematical expressions of the goal that should be predictable from existing information and models and capable of measurement from field or other data. Objectives also identify tradeoffs and constraints imposed on the decision-maker by social, legal and cultural contexts. Goals and objectives thus establish the baseline questions, directions, contexts and evaluation criteria for decisions. Establishing goals and objectives is also a critical stage in which to establish participation and create multidisciplinary and representative social and technical collaborations on which decision-making depends (Table 1). Although the mechanisms and basics for collaboration and participation in decision-making processes are beyond the scope of this book, Wengert (1976) and Lawrence *et al.* (1997) provide an introduction to the theory and practice of participation, expanding the social psychological field of procedural justice and balancing involvement in the process of decision-making with concern for decisions and their outcomes.

TECHNOLOGICAL AND METHODOLOGICAL DEVELOPMENTS

The information needs of land use issues, and purposes of decision support systems generally, provide an attractive context for application of spatial information. They also provide a driver for technological and methodological developments in spatial information handling. Many of the opportunities that land use management and planning provide for guiding development are identified and exemplified in the preceding chapters. Some general areas of development are described further here.

Description of Land Resources

Accurate and up-to-date information is needed for land management (Figure 1: Stage 3) and as input to analyses and models (Figure 1: Stage 4). The role of remote sensing in provision of up-to-date data is particularly important since the synoptic coverage of many of the existing and planned sensors provide the wide geographic coverage necessary for national and regional scale analyses. Land cover, in particular, is a very commonly used baseline dataset that provides an inventory of available resources and their distribution. Land cover also acts as an input to many environmental models and sequences of land cover descriptions through time identify and record changes in the landscape.

Several chapters identify a specific role for remote sensing either directly or indirectly in relation to land cover. In Chapter 7, Major describes developments in real-time decision support for agriculture. In Chapter 8, Ustin and Costick review the characteristics of existing and projected sensors and identify uncertainties associated with understanding the appropriate spectral and spatial scales for monitoring structural and functional states of ecosystems. In Chapter 9, Hill describes the direct and indirect use of remote sensing sources to provide quantitative and qualitative measures of pasture and grassland properties. In Chapter 10 Goodenough describes the data that space and airborne remote sensing systems provide from which to extract information about forests. These include measurement of the chemistry of forest canopies, the amount of biomass, distribution of forest species, and forest health. In Chapter 3, Dubayah et al. describe the use of Lidar to measure three-dimensional characteristics of the land surface, including vegetation and topography. These data literally provide a new dimension to land surface description and the fusion of these data with those measured using a more traditional two dimensional sensing of the environment will expand the quality and richness of resource description.

In this context, the flexible database approach to description and management of land cover characteristics, described in Chapter 2 by Loveland, is of particular value. The majority of land cover datasets used to date have been developed as a single layer as if they are to be paper maps. These data frequently have the characteristics of paper maps, even if presented within a digital environment: they are a static description exhibiting planar-enforcement and with a relatively inflexible key. The approach developed in Chapter 2 describes land cover as a suite of quantitative and qualitative characteristics using a database. A database approach provides a mechanism for storing multiple descriptions, presents an opportunity to integrate many of the different descriptors of land cover from the diverse sensors now available, and develops a data resource that is potentially of relevance to many applications. Different land cover descriptions are commonly developed for a given geographic area depending on the specific uses of the data, the sources of the data, and on the background and perspective of the dataset developers and users. Incorporating land cover descriptions into a database provides flexibility to incorporate these alternative descriptions. In many ways soil and geology data already contain these characters of complexity and flexibility in description of resources although the use of paper maps and printed monographs has limited the ability fully to exploit the detailed and rich descriptions of resources they contain. Developing digital database environments for these basic descriptors of resources will be a profitable task in linking spatial information with land use issues.

Opportunities are also offered by new remote sensing technologies that are becoming more widely accessible. Ustin and Costick discuss hyperspectral and hyperspatial remote sensing systems in Chapter 8. Plate 25 shows a real colour composite (top) from a 1m spatial resolution (hyperspatial) 128 spectral band (hyperspectral) instrument and the distribution of large woody debris in the river channel mapped using this sensor (bottom). The large woody debris, which plays a significant role in the hydrology of streams in the area depicted, is identified and mapped directly using the hyperspatial/hyperspectral data. The ability to extract landscape features directly from remote sensing sources offers great potential for resource description (either conventional mapping or digital database creation) and environmental system monitoring.

GIS and Modelling

The integration of models with Geographical Information Systems for land use management is discussed throughout the book, either directly as in Chapter 6 by Bennett, or indirectly through the need to couple analytical systems to provide insights into the land use question being addressed. Development of databases that describe land resources is complemented by close integration of analysis, and modelling. A general objective of the work described throughout the book is to establish or provide sets of tools that enhance both multi-disciplinary science and collaborative decision-making processes for land use planning and management.

Models are used as an expression of theory, to formalise and represent knowledge, to represent and test hypotheses, and to develop theoretical understanding of a system. Since models synthesise and formalise existing knowledge they also act as a mechanism for the application of that knowledge; this is their use as a predictive tool for land management and decision support. Land use models that link existing knowledge and understanding of processes with practical questions related to land management and planning provide opportunities for exploring alternative scenarios (Figure 1: Stage 4). They may also be the basis for synthesis to help communication between the multiple stakeholders typically involved in integrated management. Models can also help with issues of spatial and temporal scale and can be used for rigorous exploration of interactions between alternative land use options and their associated environmental and socio-economic impacts.

Currently the challenges faced in linking GIS with models for land management are due, in part, to the different technical and methodological approaches of land use modelling and spatial data handling. Land use modelling involves a variety of qualitative and quantitative models whose primary focus is the representation of processes that operate through time. In contrast, GIS methodologies rely on a digital representation of biophysical and socio-economic data that uses data models that closely reflect cartographic (map) production methods; this presents a temporally static description of spatial variation in data. The scientific issues facing linkage between land use modelling and GIS develop from these differences between process models and GIS data models and include:

1) a need to improve mechanisms for description, representation, and analysis of processes within GIS;

2) development of land use models that are sensitive to spatial processes and spatial organization of 'real world' environments;

3) facilitating interaction between different models and description of their output in

geographic space in relation to user requirements for scientific investigation and collaborative investigation of land use options; and

4)   questions of scale and scaling in process models and spatial description of resources.

The first three of these are directly linked to the issue of data models for linking GIS and modelling. In Chapter 6, Bennett *et al.* propose a solution based on a data model that represents dynamic geographical systems through on a triplet comprised of state, process, and relation (Bennett, 1997). A similar solution based in a data model is presented by Raper and Livingstone (1995) using a triplet of form, process and materials. Aspinall (1998) has applied the conceptual basis of Raper and Livingstones approach to coupled socio-economic and environmental modelling with GIS. Both Bennetts and Raper and Livingstones approaches to integrated GIS and modelling systems are based on, and implemented within, an object-oriented approach to software design. In practice, the majority of groups and individuals do not have access to the software facilities used for object-oriented GIS and modelling and these developments currently remain a research activity. The conceptual insights of the object-oriented approach is, however, translated into the development, operation, and application of existing GIS and modelling software through specification of the role of the system in the mind of a user. An object-oriented approach is particularly appropriate when it conforms to the way in which a problem is conceptualised; indeed the rigour associated with developing and applying object-oriented methods can provide considerable benefit to the process of developing GIS and modelling in land use decision-making if it accompanies stages 2 and 3 in the process described in Figure 1.

Linkage between GIS and models is exploited in several chapters for specific land use objectives. In Chapter 6, Bennett *et al.* couple existing process models, linear programming, and genetic algorithms with a commercial GIS to analyze the impact of alternative policy and management scenarios on land use, non-point pollution, the regional economy, and the ecology of a cypress tupelo swamp. In Chapter 12, Dale *et al.* present a suite of ecological models to support land management. These models are spatially-explicit and are developed for scenario evaluation. Use is also made of visualization for communication of results and linking the scientific evaluation of the GIS and models with land manager decision-making. In Chapter 13, Maxwell *et al.* develop a land use change prediction system designed to create a visualization of the result of alternative driving variables associated with historic land use change. Again, visualization is an important component of communication to potential users of the scientific output. In Chapter 14, White *et al.* use a cellular automata linked with models for climate, hydrology, demographics, and economic systems operating at a range of spatial and temporal scales, to model land use change across space and time. This model is also visual both in terms of the results (animated and mapped) and the model structure itself. Bennett *et al.* describe the additional role for computation in providing a virtual environment within which decision-makers and scientists can explore theory and evaluate competing management strategies; visualization plays an important role in use of spatial information and modelling for land use applications.

The role of models in developing alternative scenarios is also apparent with the use of visualization. Presenting alternative outcomes of land use decisions and selecting between them is integral to stage 4 of the decision making process described in Figure

1. Modelling and analysis can be integral to this process if alternatives can be realized and then communicated to the participants in the decision-making process.

Scale and Processes in Space and Time

Scale issues underpin many of the questions related to linkage between GIS and modelling and is a theme that runs throughout this book. What are the relationships between spatial and temporal scale, the map scales used for spatial description of resources that then act as data input to GIS and models, and scaling in process models?

Many existing process models use mechanistic understanding expressed in algorithms to describe the operation of processes through time. These descriptions of process typically use short time steps (*e.g.* day, hour, minute) and are also frequently aspatial. Coupling this form of process model with spatial data, for example in GIS, can produce compelling images although the influence of spatial organization of data is not considered. The different foundations of data in GIS, based in a spatial data model, and process modelling, based in a model operating with time but generally not space, is one reason for the emphasis on developing a common data model for coupling GIS and process modelling described above, in Chapter 6, and in the GIS and modelling literature (Goodchild *et al.*, 1993; 1996).

In Chapter 4, Goodchild discusses features of spatial data that influence analysis of spatial data. All features have important consequences for interpretation of results. In Chapter 12, Dale *et al.* use their suite of spatially-explicit models for wildlife to reveal results directly related to relative geographic location that would not be apparent had spatial location not been part of the analysis. The model described by White *et al.* in Chapter 14 is also explicitly spatial, the feedback from spatial interactions in the data being part of the reason that errors do not propagate wildly as multiple, complementary models are applied.

Identifying the appropriate spatial and temporal scale for analysis is a critical issue and research on scale influences in analysis is needed. Understanding scaling in the operation of processes, and the ways in which scientific understanding of processes that operate at one set of scales of time and space can be used to understand processes and patterns at other time and space scales is particularly important (Levin, 1992). Models produced at one spatial and temporal scale are also easily applied at other spatial and temporal scales through GIS even if this is not warranted. Research is needed on the validity of these scale translations. Some models are already used. Delcourt and Delcourt (1988) present a framework for describing relationships between spatial and temporal scaling for natural disturbance phenomena and questions of spatial and temporal scale can also be addressed using a structured analysis based in hierarchy theory (Allen and Starr, 1983). Application of spatial analysis to land use and land use changes will also be particularly helpful in establishing solutions to scale-related issues.

Uncertainty and Error

Uncertainty is manifest in a number of different ways in the examples presented in this book. The four types identified and described by Conroy in Chapter 11 provides a useful summary. Not all uncertainty is capable of description or calculation although from a scientific perspective statistical uncertainty (strictly, error) should be assessed where possible. Novel methods for communicating this form of uncertainty are also needed. The example presented in Chapter 14 suggests that multiscale, integrated modelling, although scientifically and computationally complex, may not propagate unconstrained errors due to feedbacks and interactions in the model. Rigorous analysis of uncertainty is required for models and provides a significant research challenge.

The inherent uncertainty contained in land use systems (stochastic uncertainty), our inability to completely specify and model the systems (structural uncertainty), and the limitations of management in relation to system control (partial controllability) provide a framework within which to evaluate decisions and their likely impact. In Chapter 5, Cook and Adams examine uncertainty from the decision-makers perspective. Uncertainty as risk avoidance or reduction is a social and economic goal for decision-makers. These types of uncertainty are more qualitative and are instructive about the limitations of modelling and analysis in decision-making processes. They also demonstrate a need to focus on the social components of uncertainty in decision-making in land use management and planning as well as the quantitative component associated with numerical analysis. Collectively, the different forms of uncertainty emphasize the importance of monitoring and feedback (Figure 1: Stage 6) to facilitate learning from decisions and the decision-making process. Adaptive resource management is a strategy and process that links scientific analysis with practical land management and that also directly addresses practical aspects of managing uncertainty. Monitoring and feedback are implicit in adaptive resource management and there is clearly a role for scientific methods for sampling in design and operation of appropriate monitoring systems to support the decision process outlined in Figure 1.

DECISION MAKING

Information Needs in Decision Processes

As decision support and other systems of analysis and evaluation are developed to contribute to land use management and planning, the study of complete sequences of decision processes, and the role of objective information in support of decisions, is needed. An objective of such studies will be to facilitate links between the objective information scientists attempt to provide and the legal, cultural, political, and socio-economic context within which this information is interpreted and used. The presentation and interpretation of scientific information also needs to be considered in these studies.

The differences in timescales over which decisions have to be made, usually relatively short, and the timescales over which scientists can produce decision support systems or construct and run models, usually relatively long, are also of concern both for successful development of decision support systems and applying models to particular land use issues. Developing modular decision support systems containing models, interfaces, and links with geographic databases will reduce the time taken to construct models and any analysis environment needed to address land use issues as they arise. This may take place on a single computer system although the Internet also has a role to play in this form of modular development. Increasingly data can readily be found and retrieved from data warehouses and other sources on the Internet. In the USA, data clearinghouses such as those of the National Spatial Data Infrastructure provide an Internet-based geospatial data library with many datasets available online in GIS format. A corresponding catalogue and toolbox of models and other analytical tools would be beneficial. Internet map server technologies may also be developed to provide appropriate analytical results through WWW browsers. This server-based analysis could also facilitate increased participation in land use management and planning through provision of data and derived information to anyone with WWW access and browser software. Although such access to information can encourage wide involvement in decision-making, it will require attention to the processes and management of electronic participation.

EVALUATION

Monitoring and Feedback Systems

Monitoring and feedback is the sixth stage of the decision process outlined in Figure 1. This provides an opportunity for constructive use of spatial information and spatial data handling technologies. Remote sensing and other sampling and observation systems can be integrated with spatial data management and analysis technologies to monitor the consequences and impacts of land use decisions. In Chapter 5, Cook and Adams describe the use of on-the-go sensors to record crop and soil attributes during routine management operations; these data support interpretation of decisions made in precision agriculture operations and provide feedback into analysis and decisions for the following season. Such an approach will also be of value for land use decisions at other spatial and temporal scales. A variety of indicators of the extent to which decisions meet the goals and objectives established in Stage 1 of the decision process could be established and their use be integral to evaluating the success of decisions and providing feedback in adaptive management strategies.

Indicators can also be used within the modelling and analysis process to evaluate alternative land use scenarios. In this case, indicators summarize a variety of possible impacts and consequences of land use changes. In Chapter 13, Maxwell *et al.* use a simple, single indicator to summarize and compare possible environmental impacts of alternative land use scenarios output by their model. This approach can be extended with more complex systems of indicators, each the output of one or more models, and used to examine potential impacts of land use changes on different features of the land use system. This will help comparison and evaluation of alternative land use scenarios.

Development in Real World Land Use Applications

Ultimately the examples and systems of data development, management, analysis and modelling described in this book need to be tested in practical, place-based real world land use management and planning applications. GIS, remote sensing, and modelling all have clear roles to play. Appreciating these roles within the social context of decision-making, and developing the use of spatial information in collaboration with varied stakeholders, will expand the relevance and application of spatial information and associated technologies.

REFERENCES

1. T.F.H. Allen and T.B. Starr (1983) *Hierarchy: Perspectives for Ecological Complexity.* (Chicago: Chicago University Press).

2. R.J. Aspinall (1998) Coupling biophysical and socio-economic models with Geographic Information Systems for integrated catchment management, in *Multiple Land Use and Catchment Management*, edited by M. Cresser and K. Pugh, pp. 109–123. (Aberdeen: MLURI).

3. J.A. Bellamy, G.T McDonald, G.J. Syme and J.E. Butterworth (1999) *Society Nat. Resour.*, **12**, 337–353.

4. D.A. Bennett (1997) *Geog. Environ. Model.*, **1**, 115–134.

5. R.L. Lawrence, S.E. Daniels and G.H. Stankey (1997) *Society Nat. Resour.*, **10**, 577–89.

6. J. Raper and D. Livingstone (1995) *Int. J. Geog. Inform. Syst.*, **9**, 359–383.

7. H.H. Shugart, G.B. Bonan, D.L. Urban, W.K. Lauenroth, W.J. Parton and G.M. Hornberger (1991) Computer models and long-term ecological research, in *Long-term Ecological Research: An International Perspective, SCOPE 47*, edited by P.G. Risser, pp. 211–239. (New York: John Wiley and Sons).

8. N. Wengert (1976) *Nat. Resour. J.*, **16**, 23–40.

INDEX

T - #0618 - 071024 - C0 - 254/178/6 - PB - 9780367578909 - Gloss Lamination